Plant Cell Electroporation
and Electrofusion Protocols

Methods in Molecular Biology™ Series

John M. Walker, SERIES EDITOR

Methods in Molecular Biology™ • 55

Plant Cell Electroporation and Electrofusion Protocols

Edited by

Jac A. Nickoloff

Harvard University, Boston, MA

Humana Press ☀ Totowa, New Jersey

© 1995 Humana Press Inc.
999 Riverview Drive, Suite 208
Totowa, New Jersey 07512

This publication is printed on acid-free paper. ∞
ANSI Z39.48-1984 (American National Standards Institute)
Permanence of Paper for Printed Library Materials

Photocopy Authorization Policy:
Authorization to photocopy items for internal or personal use, or the internal or personal use of specific clients, is granted by Humana Press Inc., provided that the base fee of US $4.00 per copy, plus US $00.20 per page, is paid directly to the Copyright Clearance Center at 222 Rosewood Drive, Danvers, MA 01923. For those organizations that have been granted a photocopy license from the CCC, a separate system of payment has been arranged and is acceptable to Humana Press Inc. The fee code for users of the Transactional Reporting Service is: [0-89603-328-7/95 $4.00 + $00.20].

Printed in the United States of America. 10 9 8 7 6 5 4 3 2 1

Library of Congress Cataloging-in-Publication Data

Plant cell electroporation and electrofusion protocols/edited by Jac A. Nickoloff.
    p.   cm—(Methods in molecular biology™;55)
    Includes indexes.
    ISBN 0-89603-328-7 (alk. paper)
    1. Plant genetic engineering—Laboratory manuals. 2. Plant genetic transformation—Laboratory manuals. 3. Crops—Genetic engineering—Laboratory manuals. 4. Electroporation—Laboratory manuals. 5. Electrofusion—Laboratory manuals. I. Nickoloff, Jac A. II. Series: Methods in molecular biology™ (Totowa, NJ); 55.
QK981.5.P55   1995
581.87'328—dc20   95-358
                                             CIP

Preface

Gene transfer is an essential technology for improving our understanding of gene structure and function. Although there are many methods by which DNA may be introduced into cells—including heat and chemical treatments, and microinjection—electroporation has been found to be the most versatile gene transfer technique. Electroporation is effective with a wide variety of cell types, including those that are difficult to transform by other means. For many cell types, electroporation is either the most efficient or the only means known to effect gene transfer. The early and broad success of electric field-mediated DNA transfer soon prompted researchers to investigate electroporation for transferring other types of molecules into cells, including RNA, enzymes, antibodies, and analytic dyes.

The first section of *Plant Cell Electroporation and Electrofusion Protocols* includes two chapters that serve as a guide to theoretical and practical aspects of electroporation, and will be of particular interest to those developing protocols for as yet untested species or cell types, and a third chapter that describes commercially available electroporation instruments. The remaining chapters describe well-tested protocols for DNA electrotransfection, electroporation of other biomolecules, or cell electrofusion. These chapters also include brief discussions of alternatives to electric field-based methods, citing the advantages and limitations of the various methods for achieving specific goals. Electroporation has become a favored method for introducing DNA into, and fusing plant cells, and as such these technologies promise to play a pivotal role in the development of more productive, hardy crop plants, and of plants resistant to insects and microbial pathogens. Of particular interest is the recently developed method for pollen electrotransformation, since this

v

holds significant promise for making stably inherited changes in plants by direct modification of the germ line. Technologies for transforming plant cells lagged behind those for transforming microorganisms and animal cells. The advent of plant cell electrotransformation methods has had a strong impact on both basic and applied plant research.

Although electroporation procedures for different cell types are often similar, subtle differences in either electrical parameters or growth conditions can have strong effects on transfection or fusion efficiencies, and such factors are often important to control when optimum efficiencies are required. Each protocol, therefore, provides considerable detail about conditions for growing and preparing cell and tissue samples, with many helpful troubleshooting tips. Although widely used, electroporation is still a relatively young technology, and it is expected that this collection of protocols will both propel the electric field technologies forward and facilitate the growth in our understanding of the biological processes that these technologies are used to explore.

I want to thank all of the contributors for their timely and high quality submissions, with special thanks to Dr. Patrick Gallois for his many recommendations of potential contributors. I also want to thank series editor Dr. John Walker for his valuable assistance and advice during the editing process.

Jac A. Nickoloff

Contents

vii

CONTENTS FOR THE COMPANION VOLUME

Electroporation Protocols for Microorganisms

CONTENTS FOR THE COMPANION VOLUME

Animal Cell Electroporation and Electrofusion Protocols

xi

Contributors

GEORGE W. BATES • *Department of Biological Science, Florida State University, Tallahassee, FL*

JIANPING CHENG • *Plant Sciences Institute, USDA, ARS, Beltsville, MD*

AMKE DEN DULK-RAS • *Institute of Molecular Plant Sciences, Clusius Laboratory, Leiden University, Leiden, The Netherlands*

KLAUS EIMERT • *Department of Genetics, University of Halle, Saale, Germany*

PATRICK GALLOIS • *Laboratoire de Physiologie et Biologie Moleculaire Vegetale, Université de Perpignan, France*

JOAN S. GEBHARDT • *Plant Molecular Biology Laboratory, USDA-ARS, Beltsville, MD*

GUNTER A. HOFMANN • *Genetronics Inc., San Diego, CA*

PAUL J. J. HOOYKAAS • *Institute of Molecular Plant Sciences, Clusius Laboratory, Leiden University, Leiden, The Netherlands*

SEK WEN HUI • *Membrane Biophysics Laboratory, Biophysics Department, Roswell Park Cancer Institute, Buffalo, NY*

LAURA R. KELLER • *Department of Biological Science, Florida State University, Tallahassee, FL*

SUSAN M. KOEHLER • *USDA, APHIS, BBEP, Hyattsville, MD*

JHY-JHU LIN • *Life Technologies, Gaithersburg, MD*

KEITH LINDSEY • *Depatment of Botany, University of Leicester, UK*

RENEE MALONE • *Waksman Institute, Rutgers University, Piscataway, NJ*

KATHLEEN A. MARRS • *Department of Biological Sciences, Stanford University, Stanford, CA*

BENJAMIN F. MATTHEWS • *Plant Molecular Biology Laboratory, USDA-ARS, Beltsville, MD*

LYNN E. MURRY • *Sandoz Agro, Inc., Palo Alto, CA*

CAROL A. RHODES • *Sandoz Agro, Inc., Palo Alto, CA*

JAMES A. SAUNDERS • *Soybean and Alfalfa Research Laboratory, Plant Sciences Institute, USDA-ARS, Beltsville, MD*

FRANK SIEGEMUND • *Department of Genetics, University of Halle, Saale, Germany*

HAROLD N. TRICK • *Department of Biological Sciences, Florida State University, Tallahassee, FL*

J. C. CARLE URIOSTE • *Department of Biological Sciences, Stanford University, Stanford, CA*

JAMES C. WEAVER • *Harvard–MIT Division of Health Sciences and Technology, Massachusetts Institute of Technology, Cambridge, MA*

LEI ZHANG • *Laboratory of Bioelectrochemistry, Institute of Molecular Biotechnology, Jena, Germany. Present Address: Department of Biology, Massachusetts Institute of Technology, Cambridge, MA*

PART I
THEORY AND INSTRUMENTATION

CHAPTER 1

Electroporation Theory

Concepts and Mechanisms

James C. Weaver

1. Introduction

Application of strong electric field pulses to cells and tissue is known to cause some type of structural rearrangement of the cell membrane. Significant progress has been made by adopting the hypothesis that some of these rearrangements consist of temporary aqueous pathways ("pores"), with the electric field playing the dual role of causing pore formation and providing a local driving force for ionic and molecular transport through the pores. Introduction of DNA into cells in vitro is now the most common application. With imagination, however, many other uses seem likely. For example, in vitro electroporation has been used to introduce into cells enzymes, antibodies, and other biochemical reagents for intracellular assays; to load larger cells preferentially with molecules in the presence of many smaller cells; to introduce particles into cells, including viruses; to kill cells purposefully under otherwise mild conditions; and to insert membrane macromolecules into the cell membrane itself. Only recently has the exploration of in vivo electroporation for use with intact tissue begun. Several possible applications have been identified, viz. combined electroporation and anticancer drugs for improved solid tumor chemotherapy, localized gene therapy, transdermal drug delivery, and noninvasive extraction of analytes for biochemical assays.

The present view is that electroporation is a universal bilayer membrane phenomenon *(1–7)*. Short (μs to ms) electric field pulses that cause

From: *Methods in Molecular Biology, Vol. 55: Plant Cell Electroporation and Electrofusion Protocols* Edited by: J. A. Nickoloff Humana Press Inc., Totowa, NJ

the transmembrane voltage, U(t), to rise to about 0.5–1.0 V cause electroporation. For isolated cells, the necessary single electric field pulse amplitude is in the range of 10^3–10^4 V/cm, with the value depending on cell size. Reversible electrical breakdown (REB) then occurs and is accompanied by greatly enhanced transport of molecules across the membrane. REB also results in a rapid membrane discharge, with U(t) returning to small values after the pulse ends. Membrane recovery is often orders of magnitude slower. Cell stress probably occurs because of relatively nonspecific chemical exchange with the extracellular environment. Whether or not the cell survives probably depends on the cell type, the extracellular medium composition, and the ratio of intra- to extracellular volume. Progress toward a mechanistic understanding has been based mainly on theoretical models involving transient aqueous pores. An electric field pulse in the extracellular medium causes the transmembrane voltage, U(t), to rise rapidly. The resulting increase in electric field energy within the membrane and ever-present thermal fluctuations combine to create and expand a heterogeneous population of pores. Scientific understanding of electroporation at the molecular level is based on the hypothesis that pores are microscopic membrane perforations, which allow hindered transport of ions and molecules across the membrane.

These pores are presently believed to be responsible for the following reasons:

1. Dramatic electrical behavior, particularly REB, during which the membrane rapidly discharges by conducting small ions (mainly Na^+ and Cl^-) through the transient pores. In this way, the membrane protects itself from destructive processes;
2. Mechanical behavior, such as rupture, a destructive phenomenon in which pulses too small or too short cause REB and lead to one or more supracritical pores, and these expand so as to remove a portion of the cell membrane; and
3. Molecular transport behavior, especially the uptake of polar molecules into the cell interior.

Both the transient pore population, and possibly a small number of metastable pores, may contribute. In the case of cells, relatively nonspecific molecular exchange between the intra- and extracellular volumes probably occurs, and can lead to chemical imbalances. Depending on the ratio of intra- and extracellular volume, the composition of the extracellular medium, and the cell type, the cell may not recover from the associated stress and will therefore die.

2. Basis of the Cell Bilayer
Membrane Barrier Function

It is widely appreciated that cells have membranes in order to separate the intra- and extracellular compartments, but what does this really mean? Some molecules utilized by cells have specific transmembrane transport mechanisms, but these are not of interest here. Instead, we consider the relatively nonspecific transport governed by diffusive permeation. In this case, the permeability of the membrane to a molecule of type "s" is $P_{m,s}$, which is governed by the relative solubility (partition coefficient), $g_{m,s}$, and the diffusion constant, $D_{m,s}$, within the membrane. In the simple case of steady-state transport, the rate of diffusive, nonspecific molecular transport, N_s, is:

$$N_s = A_m P_{m,s} \Delta C_s = A_m [g_{m,s} D_{m,s}/d] \Delta C_s \qquad (1)$$

where N_s, is the number of molecules of type "s" per unit time transported, ΔC_s is the concentration difference across the membrane, $d \approx 6$ nm is the bilayer membrane thickness, and A_m is the area of the bilayer portion of the cell membrane. As discussed below, for charged species, the small value of $g_{m,s}$ is the main source of the large barrier imposed by a bilayer membrane.

Once a molecule dissolves in the membrane, its diffusive transport is proportional to Δc_s and $D_{m,s}$. The dependence on $D_{m,s}$ gives a significant, but not tremendously rapid, decrease in molecular transport as size is increased. The key parameter is $g_{m,s}$, which governs entry of the molecule into the membrane. For electrically neutral molecules, $g_{m,s}$ decreases with molecular size, but not dramatically. In the case of charged molecules, however, entry is drastically reduced as charge is increased. The essential features of a greatly reduced $g_{m,s}$ can be understood in terms of electrostatic energy considerations.

The essence of the cell membrane is a thin (≈ 6 nm) region of low dielectric constant ($K_m \approx 2$–3) lipid, within which many important proteins reside. Fundamental physical considerations show that a thin sheet of low dielectric constant material should exclude ions and charged molecules. This exclusion is owing to a "Born energy" barrier, i.e., a significant cost in energy that accompanies movement of charge from a high dielectric medium, such as water (dielectric constant $K_w \approx 80$), into a low dielectric medium, such as the lipid interior of a bilayer membrane (dielectric constant $K_m \approx 2$) (8).

The Born energy associated with a particular system of dielectrics and charges, W_{Born}, is the electrostatic energy needed to assemble that system of dielectric materials and electric charge. W_{Born} can be computed by specifying the distribution of electrical potential and the distribution of charge, or it can be computed by specifying the electric field, E, and the permittivity $\varepsilon = K\varepsilon_0$ (K is the dielectric constant and $\varepsilon_0 = 8.85 \times 10^{-12}$ F/m) *(9)*. Using the second approach:

$$W_{Born} \equiv \int_{\substack{all\ space \\ except\ ion}} 1/2\ \varepsilon E^2 dV \tag{2}$$

The energy cost for insertion of a small ion into a membrane can now be understood by estimating the maximum change in Born energy, $\Delta W_{Born,max}$, as the ion is moved from water into the lipid interior of the membrane. It turns out that W_{Born} rises rapidly as the ion enters the membrane, and that much of the change occurs once the ion is slightly inside the low dielectric region. This means that it is reasonable to make an estimate based on treating the ion as a charged sphere of radius r_s and charge $q = ze$ with $z = \pm 1$ where $e = 1.6 \times 10^{-19}$ C. The sphere is envisioned as surrounded by water when it is located far from the membrane, and this gives ($W_{Born,i}$). When it is then moved to the center of the membrane, there is a new electrostatic energy, ($W_{Born,f}$). The difference in these two energies gives the barrier height, $\Delta W_{Born} \equiv W_{Born,f} - W_{Born,i}$. Even for small ions, such as Na^+ and Cl^-, this barrier is substantial (Fig. 1). More detailed, numerical computations confirm that ΔW_{Born} depends on both the membrane thickness, d, and ion radius, r_s.

Here we present a simple estimate of ΔW_{Born}. It is based on the recognition that if the ion diameter is small, $2r_s \approx 0.4$ nm, compared to the membrane thickness, $d \approx$ 3–6 nm, then ΔW_{Born} can be estimated by neglecting the finite size of the membrane. This is reasonable, because the largest electric field occurs near the ion, and this in turn means that the details of the membrane can be replaced with bulk lipid. The resulting estimate is:

$$\Delta W_{Born} \approx e^2/8\pi\varepsilon_0 r_s[1/K_m - 1/K_w] \approx 65\ kT \tag{3}$$

where T = 37°C = 310 K. A complex numerical computation for a thin low dielectric constant sheet immersed in water confirms this simple estimate (Fig. 1). This barrier is so large that spontaneous ion transport

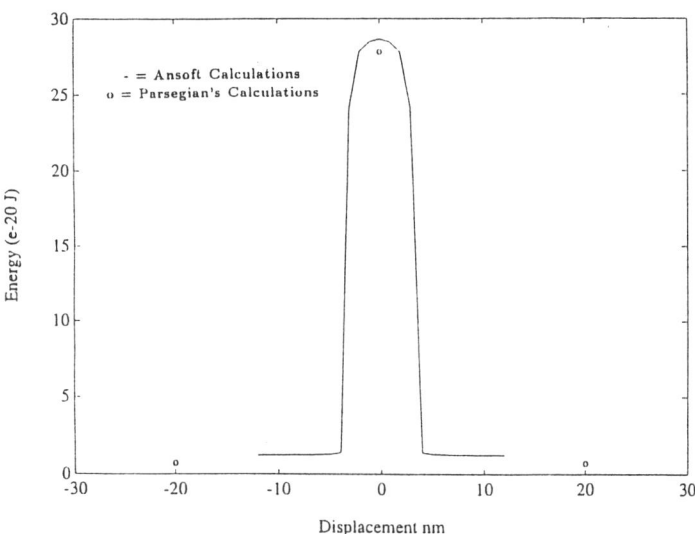

Fig. 1. Numerical calculation of the Born energy barrier for transport of a charged sphere across a membrane (thickness d = 4 nm). The numerical solution was obtained by using commercially available software (Ansoft, Inc., Pittsburgh, PA) to solve Poisson's equation for a continuum model consisting of a circular patch of a flow dielectric constant material (K_m = 2) immersed in water (K_w = 80). The ion was represented by a charged sphere of radius (r_s = 0.2 nm), and positioned at a number of different displacements on the axis of rotation of the disk. No pore was present. The electric field and the corresponding electrostatic energy were computed for each case to obtain the values plotted here as a solid line ("- Ansoft Calculations"). The single value denoted by o ("Parsegian's Calculations;" 8) is just under the Ansoft peak. As suggested by the simple estimate of Eq. (2), the barrier is large, viz. $\Delta W \approx 2.8 \times 10^{-19}$ J ≈ 65 kT. As is well appreciated, this effectively rules out significant spontaneous ion transport. The appearance of aqueous pathways ("pores"; Fig. 2) provides a large reduction in this barrier. Reproduced with permission *(47)*.

resulting from thermal fluctuations is negligible. For example, a large transmembrane voltage, U_{direct}, would be needed to force an ion directly across the membrane. The estimated value is $U_{direct} \approx 65kT/e = 1.7$ V for $z = \pm 1$. However, 1.7 V is considerably larger than the usual "resting values" of the transmembrane voltage (about 0.1 ± 0.05 V). The scientific literature on electroporation is consistent with the idea that some sort of membrane structural rearrangement occurs at a smaller voltage.

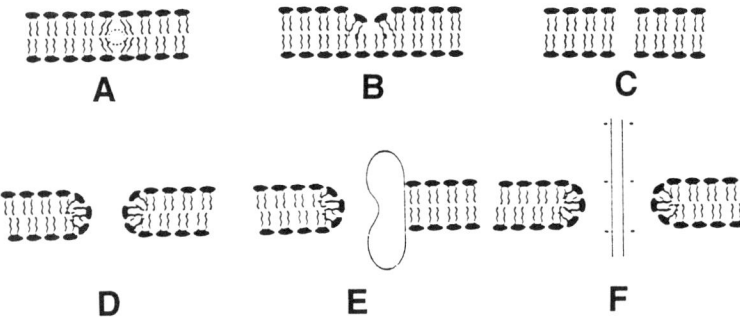

Fig. 2. Illustrations of hypothetical structures of both transient and meta-stable membrane conformations that may be involved in electroporation *(4)*. **(A)** Membrane-free volume fluctuation *(62)*, **(B)** Aqueous protrusion into the membrane ("dimple") *(12,63)*, **(C)** Hydrophobic pore first proposed as an immediate precursor to hydrophilic pores *(10)*, **(D)** Hydrophilic pore *(10,17,18)*; that is generally regarded as the "primary pore" through which ions and molecules pass, **(E)** Composite pore with one or more proteins at the pore's inner edge *(20)*, and **(F)** Composite pore with "foot-in-the-door" charged macromolecule inserted into a hydrophilic pore *(31)*. Although the actual transitions are not known, the transient aqueous pore model assumes that transitions from A → B → C or D occur with increasing frequency as U is increased. Type E may form by entry of a tethered macromolecule during the time that U is significantly elevated, and then persist after U has decayed to a small value because of pore conduction. These hypothetical structures have not been directly observed. Instead, evidence for them comes from interpretation of a variety of experiments involving electrical, optical, mechanical, and molecular transport behavior. Reproduced with permission *(4)*.

3. Aqueous Pathways ("Pores") Reduce the Membrane Barrier

A significant reduction in ΔW_{Born} occurs if the ion (1) is placed into a (mobile) aqueous cavity or (2) can pass through an aqueous channel *(8)*. Both types of structural changes have transport function based on a local aqueous environment, and can therefore be regarded as aqueous pathways. Both allow charged species to cross the membrane much more readily. Although both aqueous configurations lower ΔW_{Born}, the greater reduction is achieved by the pore *(8)*, and is the basis of the "transient aqueous pore" theory of electroporation.

Why should the hypothesis of pore formation be taken seriously? As shown in Fig. 2, it is imagined that some types of prepore structural changes can occur in a microscopic, fluctuating system, such as the bilayer membrane. Although the particular structures presented there are plausible, there is no direct evidence for them. In fact, it is unlikely that transient pores can be visualized by any present form of microscopy, because of the small size, short lifetime, and lack of a contrast-forming interaction. Instead, information regarding pores will probably be entirely indirect, mainly through their involvement in ionic and molecular transport *(4)*. Without pores, a still larger voltage would be needed to move multivalent ions directly across the membrane. For example, if $z = \pm 2$, then $U_{direct} \approx 7$ V, which for a cell membrane is huge.

Qualitatively, formation of aqueous pores is a plausible mechanism for transporting charged molecules across the bilayer membrane portion of cell membranes. The question of how pores form in a highly interactive way with the instantaneous transmembrane voltage has been one of the basic challenges in understanding electroporation.

4. Large U(t) Simultaneously Causes Increased Permeability and a Local Driving Force

Electroporation is more than an increase in membrane permeability to water-soluble species owing to the presence of pores. The temporary existence of a relatively large electric field within the pores also provides an important, local driving force for ionic and molecular transport. This is emphasized below, where it is argued that massive ionic conduction through the transient aqueous pores leads to a highly interactive membrane response. Such an approach provides an explanation of how a planar membrane can rupture at small voltages, but exhibits a protective REB at large voltages. At first this seems paradoxical, but the transient aqueous pore theory predicts that the membrane is actually protected by the rapid achievement of a large conductance. The large conductance limits the transmembrane voltage, rapidly discharges the membrane after a pulse, and thereby saves the membrane from irreversible breakdown (rupture). The local driving force is also essential to the prediction of an approximate plateau in the transport of charged molecules.

5. Membrane-Level and Cell-Level Phenomena

For applications, electroporation should be considered at two levels: (1) the membrane level, which allows consideration of both artificial and cell membranes, and (2) the cellular level, which leads to consideration of secondary processes that affect the cell. The distinction of these two levels is particularly important to the present concepts of reversible and irreversible electroporation. A key concept at the membrane level is that molecular transport occurs through a dynamic pore population. A related hypothesis is that electroporation itself can be reversible at the membrane level, but that large molecular transport can lead to significant chemical stress of a cell, and it is this secondary, cell-level event that leads to irreversible cell electroporation. This will be brought out in part of the presentation that follows.

6. Reversible and Irreversible Electroporation at the Membrane Level

Put simply, reversible electroporation involves creation of a dynamic pore population that eventually collapses, returning the membrane to its initial state of a very few pores. As will be discussed, reversible electroporation generally involves REB, which is actually a temporary high conductance state. Both artificial planar bilayer membranes and cell membranes are presently believed capable of experiencing reversible electroporation. In contrast, the question of how irreversible electroporation occurs is reasonably well understood for artificial planar bilayer membranes, but significantly more complicated for cells.

7. Electroporation in Artificial Planar and in Cell Membranes

Artificial planar bilayer membrane studies led to the first proposals of a theoretical mechanism for electroporation *(10–16)*. However, not all aspects of planar membrane electroporation are directly relevant to cell membrane electroporation. Specifically, quantitative understanding of the stochastic rupture ("irreversible breakdown") in planar membranes was the first major accomplishment of the pore hypothesis. Although cell membranes can also be damaged by electroporation, there are two possible mechanisms. The first possibility is lysis resulting from a secondary result of reversible electroporation of the cell membrane. According to this hypothesis, even though the membrane recovers (the dynamic pore population returns to the initial state), there can be so much molecular transport that the cell is chemically or osmotically stressed, and this secondary event leads to cell destruction through lysis. The second possibility is that rupture of an isolated portion of a cell membrane occurs, because one or more bounded portions of the membrane behave like small planar membranes. If this is the case, the mechanistic understanding of planar membrane rupture is relevant to cells.

8. Energy Cost to Create a Pore
at Zero Transmembrane Voltage ($U = 0$)

The first published descriptions of pore formation in bilayer membranes were based on the idea that spontaneous (thermal fluctuation driven) structural changes in the membrane could create pores. A basic premise was that the large pores could destroy a membrane by rupture, which was suggested to occur as a purely mechanical event, i.e., without electrical assistance *(17,18)*. The energy needed to make a pore was considered to involve two contributions. The first is the "edge energy," which relates to the creation of a stressed pore edge, of length $2\pi r$, so that if the "edge energy" (energy cost per length) was γ, then the cost to make the pore's edge was $2\pi r\gamma$. The second is the "area energy" change associated with removal of a circular patch of membrane, $-\pi r^2\Gamma$. Here Γ is the energy per area (both sides of the membrane) of a flat membrane.

Put simply, this process is a "cookie cutter" model for a pore creation. The free energy change, $\Delta W_p(r)$, is based on a gain in edge energy and a simultaneous reduction in area energy. The interpretation is simple: a pore-free membrane is envisioned, then a circular region is cut out of the membrane, and the difference in energy between these two states calculated, and identified as ΔW_p. The corresponding equation for the pore energy is:

$$\Delta W_p(r) = 2\pi\gamma r - \pi\Gamma r^2 \text{ at } U = 0 \qquad (4)$$

A basic consequence of this model is that $\Delta W_p(r)$ describes a parabolic barrier for pores. In its simplest form, one can imagine that pores might be first made, but then expanded at the cost of additional energy. If the barrier peak is reached, however, then pores moving over the barrier can expand indefinitely, leading to membrane rupture. In the initial models (which did not include the effect of the transmembrane voltage), spontaneous thermal fluctuations were hypothesized to create pores, but the probability of surmounting the parabolic barrier was thought to be small. For this reason, it was concluded that spontaneous rupture of a red blood cell membrane by spontaneous pore formation and expansion was concluded to be negligible *(17)*. At essentially the same time, it was independently suggested that pores might provide sites in the membrane where spontaneous translocation of membrane lipid molecules ("flip flop") should preferentially occur *(18)*.

9. Energy Cost to Create a Pore at U > 0

In order to represent the electrical interaction, a pore is regarded as having an energy associated with the change of its specific capacitance, C_p. This was first presented in a series of seven back-to-back papers *(10–16)*. Early on, it was recognized that it was unfavorable for ions to enter small pores because of the Born energy change discussed previously. For this reason, a relatively small number of ions will be available within small pores to contribute to the electrical conductance of the pore. With this justification, a pore is represented by a water-filled, rather than electrolyte-filled, capacitor. However, for small hydrophilic pores, even if bulk electrolyte exists within the pores, the permittivity would be $\varepsilon \approx 70\varepsilon_0$, only about 10% different from that of pure water.

In this case, the pore resistance is still large, $R_p = \rho_e h / \pi r^2$, and is also large in comparison to the spreading resistance discussed below. If so, the voltage across the pore is approximately U. With this in mind, in the presence of a transmembrane electric field, the free energy of pore formation should be *(10)*:

$$\Delta W_p(r,U) = 2\pi\gamma r - \pi\Gamma r^2 - 0.5C_p U^2 r^2 \qquad (5)$$

Here U is the transmembrane voltage spatially averaged over the membrane. A basic feature is already apparent in the above equation: as *U* increases, the pore energy, ΔW_p, decreases, and it becomes much more favorable to create pores. In later versions of the transient aqueous pore model, the smaller, local transmembrane voltage, U_p, for a conducting pore is used. As water replaces lipid to make a pore, the capacitance of the membrane increases slightly.

10. Heterogeneous Distribution of Pore Sizes

A spread in pore sizes is fundamentally expected *(19–22)*. The origin of this size heterogeneity is the participation of thermal fluctuations along with electric field energy within the membrane in making pores. The basic idea is that these fluctuations spread out the pore population as pores expand against the barrier described by $\Delta W_p(r,U)$. Two extreme cases illustrate this point: (1) occasional escape of large pores over the barrier described by $\Delta W_p(r,U)$ leads to rupture, and (2) the rapid creation of many small pores ($r \approx r_{min}$) causes the large conductance that is responsible for REB. In this sense, rupture is a large-pore phenomenon, and REB is a small-pore phenomenon. The moderate value of U(t) associated with rupture leads to only a modest conductance, so that there is ample time for the pore population to evolve such that one or a small number of large pores appear and diffusively pass over the barrier, which is still fairly large. The pore population associated with REB is quite different; at larger voltages, a great many more small pores appear, and these discharge the membrane before the pore population evolves any large "critical" pores that lead to rupture.

11. Quantitative Explanation of Rupture

As the transmembrane voltage increases, the barrier $\Delta W_p(r, U)$ changes its height, ΔW_{max} and the location of its peak. The latter is associated with a critical pore radius, r_c, such that pores with $r > r_c$ tend to expand without limit. A property of $\Delta W_p(r, U)$ is that both ΔW_{max} and r_c decrease as U increases. This provides a readily visualized explanation of planar membrane rupture: as U increases, the barrier height decreases, and this increases the probability of the membrane acquiring one or more pores with $r > r(U)_c$. The appearance of even one supracritical pore is, however, sufficient to rupture the membrane. Any pore with $r > r_c$ tends to expand until it reaches the macroscopic aperture that defines the planar membrane. When this occurs, the membrane material has all collected at the aperture, and it makes no sense to talk about a membrane being present. In this case, the membrane is destroyed.

The critical pore radius, r_c, associated with the barrier maximum, $\Delta W_{p,max} = \Delta W_p(r_c, U)$, is *(10)*:

$$r_c = (\gamma/\Gamma + 0.5C_pU^2) \text{ and } \Delta W_{p,max} = \pi\gamma^2/\Gamma + 0.5C_pU^2) \qquad (6)$$

The associated pore energy, $\Delta W_{p,max}$ also decreases. Overcoming energy barriers generally depends nonlinearly on parameters, such as *U,* because Boltzmann factors are involved. For this reason, a nonlinear dependence on *U* was expected.

The electrical conductance of the membrane increases tremendously because of the appearance of pores, but the pores, particularly the many small ones, are not very good conductors. The reason for this relatively poor conduction of ions by small pores is again the Born energy change; conduction within a pore can be suppressed over bulk electrolyte conduction because of Born energy exclusion owing to the nearby low dielectric constant lipid. The motion of ions through a pore only somewhat larger than the ion itself can be sterically hindered. This has been accounted for by using the Renkin equation to describe the essential features of hindrance *(23)*. This function provides for reduced transport of a spherical ion or molecule of radius r_s through cylindrical pathway of radius r (representing a pore) *(20,21,24)*.

12. Planar Membrane Destruction by Emergence of Even One "Critical Pore"

As a striking example of the significance of heterogeneity within the pore population, it has been shown that one or a small number of large pores can destroy the membrane by causing rupture *(11)*. The original approach treated the diffusive escape of pores over an energy barrier. Later, an alternative, simpler approach for theoretically estimating the average membrane lifetime against rupture, $\bar{\tau}$, was proposed *(25)*. This approach used an absolute rate estimate for critical pore appearance in which a Boltzmann factor containing $\Delta W_p/kT$ and an order of magnitude estimate for the prefactor was used. The resulting estimate for the rate of critical pore appearance is:

$$\bar{\tau} \approx (1/\nu_0 V_m) \exp (+\Delta W_{p,c}/kT) \tag{7}$$

This estimate used an attempt rate density, ν_0, which is based on a collision frequency density within the fluid bilayer membrane. The order of magnitude of ν_0 was obtained by estimating the volume density of collisions per time in the fluid membrane. The factor $V_m = hA_m$ is the total volume of the membrane. By choosing a plausible value (e.g., 1 s), the value of $\Delta W_{p,c}$, and hence of U_c, can be found. This is interpreted as the critical voltage for rupture. Because of the strong nonlinear behavior of Eq. (7), using values, such as 0.1 or 10 s, results in only small differences in the predicted $U_c \approx 0.3$–0.5 V.

13. Behavior of the Transmembrane Voltage During Rupture

Using this approach, reasonable (but not perfect) agreement for the behavior of $U(t)$ was found. Both the experimental and theoretical behaviors of U exhibit a sigmoidal decay during rupture, but the duration of the decay phase is longer for the experimental values. Both are much longer than the rapid discharge found for REB. Many experiments have shown that both artificial planar bilayer membranes and cell membranes exhibit REB, and its occurrence coincides with tremendously enhanced molecular transport across cell membranes. However, the term "breakdown" is misleading, because REB is now believed to be a protective behavior, in which the membrane acquires a very large conductance in the form of pores. In planar membranes challenged by short pulses (the "charge injection" method mentioned above), a characteristic of REB is the progressively faster membrane discharge as larger and larger pulses are used *(26)*.

14. Reversible Electroporation

Unlike reversible electroporation (rupture) of planar membranes, in which the role of one or a small number of critical pores is dominant, reversible electroporation is believed to involve the rapid creation of so many small pores that membrane discharge occurs before any critical pores can evolve from the small pores. The transition in a planar membrane from rupture to REB can be qualitatively understood in terms of a competition between the kinetics of pore creation and of pore expansion. If only a few pores are present owing to a modest voltage pulse, the membrane discharges very slowly (e.g., ms) and there is time for evolution of critical pores. If a very large number of pores are present because of a large pulse, then the high conductance of these pores discharges the membrane rapidly, before rupture can occur. One basic challenge in a mechanistic understanding is to find a quantitative description of the transition from rupture to REB, i.e., to show that a planar membrane can experience rupture for modest pulses, but makes a transition to REB as the pulse amplitude is increased *(19–22)*. This requires a physical model for both pore creation and destruction, and also the behavior of a dynamic, heterogeneous pore population.

15. Conducting Pores Slow Their Growth

An important aspect of the interaction of conducting pores with the changing transmembrane voltage is that pores experience a progressively smaller expanding force as they expand *(21,27)*. This occurs because there are inhomogeneous electric fields (and an associated "spreading resistance") just outside a pore's entrance and exit, such that as the pore grows, a progressively greater fraction of U appears across this spreading resistance. This means that less voltage appears across the pore itself, and therefore, the electrical expanding pressure is less. For this reason, pores tend to slow their growth as they expand. The resistance of the internal portion of the pore is also important, and as already mentioned, has a reduced internal resistance because $\sigma_p < \sigma_e$ because of Born energy "repulsion." The voltage divider effect means simply that the voltage across the pore is reduced to:

$$U_p = U\ [R_p/(R_p + R_s)] \leq U \tag{8}$$

Here R_p is the electrical resistance associated with the pore interior, and R_s is the resistance associated with the external inhomogeneous electric field near the entrance and exit to the pore. The fact that U_p becomes less than U means that the electrical expanding force owing to the gradient of ΔW_p in pore radius space is reduced. In turn, this means that pores grow more slowly as they become larger, a basic pore response that contributes to reversibility *(21,27)*.

16. Reversible Electroporation and "Reversible Electrical Breakdown"

For planar membranes, the transition from irreversible behavior ("rupture") to reversible behavior ("REB" or incomplete reversible electrical breakdown) can be explained by the evolution of a dynamic, heterogeneous pore population *(20–22,24)*. One prediction of the transient aqueous pore model is that a planar membrane should also exhibit incomplete reversible electrical breakdown, i.e., a rapid discharge that does not bring *U* down to zero. Indeed, this is predicted to occur for somewhat smaller pulses than those that produce REB. Qualitatively, the following is believed to occur. During the initial rapid discharge, pores rapidly shrink and some disappear. As a result, the membrane conductance, G(t), rapidly reaches such a small value that further discharge occurs very slowly. On the time scale (μs) of the experiment, discharge appears to stop, and the membrane has a small transmembrane voltage, e.g., $U \approx 50$ mV.

Although irreversible electroporation of planar membranes now seems to be reasonably accounted for by a transient aqueous pore theory, the case of irreversibility in cells is more complicated and still not fully understood. The rupture of planar membranes is explained by recognizing that expansion of one or more supracritical pores can destroy the membrane. When it is created, the planar membrane covers a macroscopic aperture, but also connects to a meniscus at the edge of the aperture. This meniscus also contains phospholipids, and can be thought of as a reservoir that can exchange phospholipid molecules with the thinner bilayer membrane. As a result of this connection to the meniscus, the bilayer membrane has a total surface tension (both sides of the membrane), Γ, which favors expansion of pores. Thus, during rupture, the membrane material is carried by pore expansion into the meniscus, and the membrane itself vanishes.

However, there is no corresponding reservoir of membrane molecules in the case of the closed membrane of a vesicle or cell. For this reason, if the osmotic pressure difference across the cell membrane is zero, the cell membrane effectively has $\Gamma = 0$. For this reason, a simple vesicle cannot rupture *(28)*. Although a cell membrane has the same topology as a vesicle, the cell membrane is much more complicated, and usually contains other, membrane-connecting structures. With this in mind, suppose that a portion of a cell membrane is bounded by the cytoskeleton or some other cellular structure, such that membrane molecules can accumulate there if pores are created (Fig. 2). If so, these bounded portions of the cell membrane may be able to rupture, since a portion of the cell membrane would behave like a microscopic planar bilayer membrane. This localized but limited rupture would create an essentially permanent hole in the cell membrane, and would lead to cell death. Another possibility is that reversible electroporation occurs, with REB and a large, relatively nonspecific molecular transport (*see* Section 21.) across the cell membrane.

17. Tremendous Increase
in Membrane Conductance, G(t) During REB

Creation of aqueous pathways across the membrane is, of course, the phenomenon of interest. This is represented by the total membrane conductance, $G(t) = 1/R(t)$. As pores appear during reversible electroporation, R changes by orders of magnitude. A series of electrical experiments using a planar bilayer membrane provided conditions and results that motivated the choice of particular parameters, including the use of a very short (0.4 µs) square pulse *(26)*. In these experiments, a current pulse of amplitude I_i passes through R_N, thereby creating a voltage pulse, V_0 (Fig. 2). For $0 < t < t_{pulse}$ current flows into and/or across the membrane, and at $t = t_{pulse}$, the pulse is terminated by opening the switch. Because the generator is then electronically disconnected, membrane discharge can occur only through the membrane for a planar membrane (not true for a cell). Predictions of electroporation behavior were obtained by generating self-consistent numerical solutions to these equations.

18. Evidence for Metastable Pores

Pores do not necessarily disappear when U returns to small values. For example, electrical experiments with artificial planar bilayer membranes have shown that small pores remain after U is decreased. Other experiments with cells have examined the response of cells to dyes supplied after electrical pulsing, and find that a subpopulation of cells takes up these molecules *(29,30)*. Although not yet understood quantitatively in terms of an underlying mechanism, it is qualitatively plausible that some type of complex, metastable pores can form. Such pores may involve other components of a cell, e.g., the cytoskeleton or tethered cytoplasmic molecules (Fig. 2), that lead to metastable pores. For example, entry of a portion of a tethered, charged molecule should lead to a "foot-in-the-door" mechanism in which the pore cannot close *(31)*. However, pore destruction is not well understood. Initial theories assumed that pore disappearance occurs independently of other pores. This is plausible, since pores are widely spaced even when the total (aqueous) area is maximum *(22)*. Although this approximate treatment has contributed to reasonable theoretical descriptions of some experimental behavior, a complete, detailed treatment of pore disappearance remains an unsolved problem.

19. Interaction of the Membrane with the External Environment

It is not sufficient to describe only the membrane. Instead, an attempt to describe an experiment should include that part of the experimental apparatus that directly interacts with the membrane. Specifically, the electrical properties of the bathing electrolyte, electrodes, and output characteristics of the pulse generator should be included. Otherwise, there is no possibility for including the limiting effects of this part of the experiment. Clearly there is a pathway by which current flows in order to cause interfacial polarization, and thereby increase $U(t)$.

An initial attempt to include membrane–environment interactions used a simple circuit model to represent the most important aspects of the membrane and the external environment, which shows the relationship among the pulse generator, the charging pathway resistance, and the membrane *(19,21)*. The membrane is represented as the membrane capacitance, C, connected in parallel with the membrane resistance, $R(t)$. As pores begin to appear in the membrane, the membrane conductance $G(t) = 1/R(t)$ starts to increase, and therefore $R(t)$ drops. The membrane does not experience the applied pulse immediately, however, since the membrane capacitance has to charge through the external resistance of the electrolyte, which baths the membrane, the electrode resistance, and the output resistance of the pulse generator. This limitation is represented by a single resistor, R_E. This explicit, but approximate, treatment of the membrane's environment provides a reasonable approach to achieving theoretical descriptions of measurable quantities that can be compared to experimental results.

20. Fractional Aqueous Area
of the Membrane During Electroporation

The membrane capacitance is treated as being constant, which is consistent with experimental data *(32)*. It is also consistent with the theoretical model, as shown by computer simulations that use the model to predict correctly basic features of the transmembrane voltage, $U(t)$. The simulation allows the slight change in C to be predicted simultaneously, and finds that only a small fraction ($F_{w,\max} \approx 5 \times 10^{-4}$) of the membrane becomes aqueous through the appearance of pores. The additional capacitance owing to this small amount of water leads to a slight (on the order of 1%) change in the capacitance *(22)*, which is consistent with experimental results *(32)*.

The fractional aqueous area, $F_w(t)$, changes rapidly with time as pores appear, but is predicted to be less than about 0.1% of the membrane, even though tremendous increases in ionic conduction and molecular transport take place. This is in reasonable agreement with experimental findings. According to present understanding, the minimum pore size is $r_{\min} \approx 1$ nm, which means that the small ions that comprise physiologic saline can be conducted. For larger or more charged species, however, the available fractional aqueous area, $F_{w,s}$, is expected to decrease. This is a consequence of a heterogeneous pore population. With increasing molecular size and/or charge, fewer and fewer pores should participate, and this means that $F_{w,s}$ should decrease as the size and charge of "s" increase.

21. Molecular Transport Owing
to Reversible Electroporation

Tremendously increased molecular transport *(33,34)* is probably the most important result of electroporation for biological research (Table 1). Although clearly only partially understood, much of the evidence to date supports the view that electrophoretic transport through pores is the major mechanism for transport of charged molecules *(20,24,35,36)*.

Table 1
Candidate Mechanisms for Molecular Transport Through Pores *(20)[a]*

Mechanism	Molecular basis
Drift	Velocity in response to a local physical (e.g., electrical) field
Diffusion	Microscopic random walk
Convection	Fluid flow carrying dissolved molecules

[a]The dynamic pore population of electroporation is expected to provide aqueous pathways for molecular transport. Water-soluble molecules should be transported through the pores that are large enough to accommodate them, but with some hindrance. Although not yet well established, electrical drift may be the primary mechanism for charged molecules *(20–35)*.

One surprising observation is the molecular transport caused by a single exponential pulse can exhibit a plateau, i.e., transport becomes independent of field pulse magnitude, even though the net molecular transport results in uptake that is far below the equilibrium value $N_s = V_{cell}c_{ext}$ *(37–40)*. Here N_s is the number of molecules taken up by a single cell, V_{cell} is the cell volume, and c_{ext} is the extracellular concentration in a large volume of pulsing solution.

A plateauing of uptake that is independent of equilibrium uptake $(\overline{n}_s = V_{cell}c_{s,ext})$ may be a fundamental attribute of electroporation. Initial results from a transient aqueous pore model show that the transmembrane voltage achieves an almost constant value for much of the time during an exponential pulse. If the local driving force is therefore almost constant, the transport of small charged molecules through the pores may account for an approximate plateau *(24)*. Transport of larger molecules may require deformation of the pores, but the approximate constancy of $U(t)$ should still occur, since the electrical behavior is dominated by the many smaller pores. These partial successes of a transient aqueous pore theory are encouraging, but a full understanding of electroporative molecular transport is still to be achieved.

22. Terminology and Concepts: Breakdown and Electropermeabilization

Based on the success of the transient aqueous pore models in providing reasonably good quantitative descriptions of several key features of electroporation, the existence of pores should be regarded as an attractive hypothesis (Table 2). With this in mind, two widely used terms, "breakdown" and "electropermeabilization," should be re-examined. First, "breakdown" in the sense of classic dielectric breakdown is mis-

Table 2
Successes of the Transient Aqueous Pore Model[a]

Behavior	Pore theory accomplishment
Stochastic nature of rupture	Explained by diffusive escape of very large pores (10)
Rupture voltage, U_c	Average value reasonably predicted (10,25,64)
Reversible electrical breakdown	Transition from rupture to REB correctly predicted (21)
Fractional aqueous area	$F_{w,ions} \leq 10^{-3}$ predicted; membrane conductance agrees (22)
Small change in capacitance	Predicted to be <2% for reversible electroporation (22)
Plateau in charged molecule transport	Approximate plateau predicted for exponential pulses (24)

[a]Successful predictions of the transient aqueous pore model for electroporation at the present time. These more specific descriptions are not accounted for simply by an increased permeability or an ionizing type of dielectric breakdown. The initial, combined theoretical and experimental studies convincingly showed that irreversible breakdown ("rupture") was not the result of a deterministic mechanism, such as compression of the entire membrane, but could instead be quantitatively accounted for by transient aqueous pores (10). Recent observations of charged molecule uptake by cells that exhibits a plateau, but is far below the equilibrium value cannot readily be accounted for by any simple, long-lasting membrane permeability increase, but is predicted by the transient aqueous pore model.

leading. After all, the maximum energy available to a monovalent ion or molecule for $U \approx 0.5–1$ V is only about one-half to 1 ev. This is too small to ionize most molecules, and therefore cannot lead to conventional avalanche breakdown in which ion pairs are formed (41). Instead, a better term would be "high conductance state," since it is the rapid membrane rearrangement to form conducting aqueous pathways that discharges the membrane under biochemically mild conditions (42). Second, in the case of electropermeabilization, "permeabilization" implies only that a state of increased permeability has been obtained. This phenomenological term is directly relevant only to transport. It does not lead to the concept of a stochastic membrane destruction, the idea of "reversible electrical breakdown" as a protective process in the transition from rupture to REB, or the plateau in molecular transport for small charged molecules. Thus, although electroporation clearly causes an increase in permeability, electroporation is much more, and the abovementioned additional features cannot be explained solely by an increase in permeability.

23. Membrane Recovery

Recovery of the membrane after pulsing is clearly essential to achieving reversible behavior. Presently, however, relatively little is known about the kinetics of membrane recovery after the membrane has been discharged by REB. Some studies have used "delayed addition" of molecules to determine the integrity of cell membranes at different times after pulsing. Such experiments suggest that a subpopulation of cells occurs that has delayed membrane recovery, as these cells are able to take up molecules after the pulse. In addition to "natural recovery" of cell membranes, the introduction of certain surfactants has been found to accelerate membrane recovery, or at least re-establishment of the barrier function of the membrane *(43)*. Accelerated membrane recovery may have implications for medical therapies for electrical shock injury, and may also help us to understand the mechanism by which membranes recover.

24. Cell Stress and Viability

Complete cell viability, not just membrane recovery, is usually important to biological applications of electroporation, but in the case of electroporation, determination of cell death following electroporation is nontrivial. After all, by definition, electroporation alters the permeability of the membrane. This means that membrane-based short-term tests (vital stains, membrane exclusion probes) are therefore not necessarily valid *(29)*. If, however, the cells in question can be cultured, assays based on clonal growth should provide the most stringent test, and this can be carried out relatively rapidly if microcolony (2–8 cells) formation is assessed *(44)*. This was done using microencapsulated cells. The cells are initially incorporated into agarose gel microdrops (GMDs), electrically pulsed to cause electroporation, cultured while in the microscopic (e.g., 40–100 μm diameter) GMDs, and then analyzed by flow cytometry so that the subpopulation of viable cells can be determined *(45,46)*.

Cellular stress caused by electroporation may also lead to cell death without irreversible electroporation itself having occurred. According to our present understanding of electroporation itself, both reversible and irreversible electroporation result in transient openings (pores) of the membrane. These pores are often large enough that molecular transport is expected to be relatively nonspecific. As already noted, for irreversible electroporation, it is plausible that a portion of the cell membrane behaves much like a small planar membrane, and therefore can undergo rupture. In the case of reversible electroporation, significant molecular transport between the intra- and extracellular volumes may lead to a significant chemical imbalance. If this imbalance is too large, recovery may not occur, with cell death being the result. Here it is hypothesized that the volumetric ratio:

$$R_{vol} \equiv (V_{extracellular}/V_{intracellular}) \tag{9}$$

may correlate with cell death or survival *(47)*. According to this hypothesis, for a given cell type and extracellular medium composition, $R_{vol} \gg$ 1 (typical of in vitro conditions, such as cell suspensions and anchorage-dependent cell culture) should favor cell death, whereas the other extreme $R_{vol} \ll 1$ (typical of in vivo tissue conditions) should favor cell survival. If correct, for the same degree of electroporation, significantly less damage may occur in tissue than in body fluids or under most in vitro conditions.

25. Tissue Electroporation

Tissue electroporation is a relatively new extension of single-cell electroporation under in vitro conditions, and is of interest because of possible medical applications, such as cancer tumor therapy *(48–50)*, transdermal drug delivery *(51,52)*, noninvasive transdermal chemical sensing *(4)*, and localized gene therapy *(53,54)*. It is also of interest because of its role in electrical injury *(43,55,56)*. The interest in tissue electroporation is growing rapidly, and may lead to many new medical applications. The basic concept is that application of electric field pulses to tissue generally results in a localized, large electric field developing across the lipid-based barriers within the tissue. This can result in the creation of new aqueous pathways across the barrier, just where they are needed in order to achieve local drug delivery. Relevant barriers are not only the single bilayer membranes of cells, but one or more tissue monolayers in which cells are connected by tight junctions (essentially two bilayers in series per monolayer), and the stratum corneum of the skin, which can be regarded very approximately as about 100 bilayer membranes in series. In such cases, it is envisioned that electroporation is to be used with living human subjects. With this in mind, it is significant that several studies support the view that electroporation conditions can be found that result in negligible damage, both in isolated cells *(57–59)* and in intact tissue in vivo *(60,61)*. Increased use of electroporation for drug delivery implies that a much better mechanistic understanding of electroporation will be needed to secure both scientific and regulatory acceptance.

26. Summary

The basic features of electrical and mechanical behavior of electroporated cell membranes are reasonably well established experimentally. Overall, the electrical and mechanical features of electroporation are consistent with a transient aqueous pore hypothesis, and several features, such as membrane rupture and reversible electrical breakdown, are reasonably well described quantitatively. This gives confidence that "electroporation" is an attractive hypothesis, and that the appearance of temporary pores owing to the simultaneous contributions of thermal fluctuations ("kT energy") and an elevated transmembrane voltage ("electric field energy") is the microscopic basis of electroporation.

Acknowledgments

I thank J. Zahn, T. E. Vaughan, M. A. Wang, R. M. Prausnitz, R. O. Potts, U. Pliquett, J. Lin, R. Langer, L. Hui, E. A. Gift, S. A. Freeman, Y. Chizmadzhev, and V. G. Bose for many stimulating and critical discussions. This work supported by NIH Grant GM34077, Army Research Office Grant No. DAAL03-90-G-0218, NIH Grant ES06010, and a computer equipment grant from Stadwerke Düsseldorf, Düsseldorf, Germany.

References

1. Neumann, E., Sowers, A., and Jordan, C. (eds.) (1989) *Electroporation and Electrofusion in Cell Biology.* Plenum, New York.
2. Tsong, T. Y. (1991) Electroporation of cell membranes. *Biophys. J.* **60,** 297–306.
3. Chang, D. C., Chassy, B. M., Saunders, J. A., and Sowers, A. E. (eds.) (1992) *Guide to Electroporation and Electrofusion.* Academic.
4. Weaver, J. C. (1993) Electroporation: a general phenomenon for manipulating cells and tissue. *J. Cell. Biochem.* **51,** 426–435.
5. Orlowski, S. and Mir, L. M. (1993) Cell electropermeabilization: a new tool for biochemical and pharmacological studies. *Biochim. Biophys. Acta* **1154,** 51–63.
6. Weaver, J. C. (1994) Electroporation in cells and tissues: a biophysical phenomenon due to electromagnetic fields. *Radio Sci.* (in press).
7. Weaver, J. C. and Chizmadzhev, Y. A. Electroporation, in *CRC Handbook of Biological Effects of Electromagnetic Fields,* 2nd ed. (Polk, C. and Postow, E., eds.), CRC, Boca Raton (submitted).
8. Parsegian, V. A. (1969) Energy of an ion crossing a low dielectric membrane: solutions to four relevant electrostatic problems. *Nature* **221,** 844–846.
9. Zahn, M. (1979) *Electromagnetic Field Theory: A Problems Solving Approach,* Wiley, New York.
10. Abidor, I. G., Arakelyan, V. B., Chernomordik, L. V., Chizmadzhev, Yu. A., Pastushenko, V. F., and Tarasevich, M. R. (1979) Electric breakdown of bilayer

membranes: I. The main experimental facts and their qualitative discussion. *Bioelectrochem. Bioenerg.* **6**, 37–52.

11. Pastushenko, V. F., Chizmadzhev, Yu. A., and Arakelyan, V. B. (1979) Electric breakdown of bilayer membranes: II. Calculation of the membrane lifetime in the steady-state diffusion approximation. *Bioelectrochem. Bioenerg.* **6**, 53–62.

12. Chizmadzhev, Yu. A., Arakelyan, V. B., and Pastushenko, V. F. (1979) Electric breakdown of bilayer membranes: III. Analysis of possible mechanisms of defect origin. *Bioelectrochem. Bioenerg.* **6**, 63–70.

13. Pastushenko, V. F., Chizmadzhev, Yu. A., and Arakelyan, V. B. (1979) Electric breakdown of bilayer membranes: IV. Consideration of the kinetic stage in the case of the single-defect membrane. *Bioelectrochem. Bioenerg.* **6**, 71–79.

14. Arakelyan, V. B., Chizmadzhev, Yu. A., and Pastushenko, V. F. (1979) Electric breakdown of bilayer membranes: V. Consideration of the kinetic stage in the case of the membrane containing an arbitrary number of defects. *Bioelectrochem. Bioenerg.* **6**, 81–87.

15. Pastushenko, V. F., Arakelyan, V. B., and Chizmadzhev, Yu. A. (1979) Electric breakdown of bilayer membranes: VI. A stochastic theory taking into account the processes of defect formation and death: membrane lifetime distribution function. *Bioelectrochem. Bioenerg.* **6**, 89–95.

16. Pastushenko, V. F., Arakelyan, V. B., and Chizmadzhev, Yu. A. (1979) Electric breakdown of bilayer membranes: VII. A stochastic theory taking into account the processes of defect formation and death: statistical properties. *Bioelectrochem. Bioenerg.* **6**, 97–104.

17. Litster, J. D. (1975) Stability of lipid bilayers and red blood cell membranes. *Phys. Lett.* **53A**, 193,194.

18. Taupin, C., Dvolaitzky, M., and Sauterey, C. (1975) Osmotic pressure induced pores in phospholipid vesicles. *Biochemistry* **14**, 4771–4775.

19. Powell, K. T., Derrick, E. G., and Weaver, J. C. (1986) A quantitative theory of reversible electrical breakdown. *Bioelectrochem. Bioelectroenerg.* **15**, 243–255.

20. Weaver, J. C. and Barnett, A. (1992) Progress towards a theoretical model of electroporation mechanism: membrane electrical behavior and molecular transport, in *Guide to Electroporation and Electrofusion* (Chang, D. C., Chassy, B. M., Saunders, J. A., and Sowers, A. E., eds.), Academic.

21. Barnett, A. and Weaver, J. C. (1991) Electroporation: a unified, quantitative theory of reversible electrical breakdown and rupture. *Bioelectrochem. Bioenerg.* **25**, 163–182.

22. Freeman, S. A., Wang, M. A., and Weaver, J. C. (1994) Theory of electroporation for a planar bilayer membrane: predictions of the fractional aqueous area, change in capacitance and pore-pore separation. *Biophysical J.* **67**, 42–56.

23. Renkin, E. M. (1954) Filtration, diffusion and molecular sieving through porous cellulose membranes. *J. Gen. Physiol.* **38**, 225–243.

24. Wang, M. A., Freeman, S. A., Bose, V. G., Dyer, S., and Weaver, J. C. (1993) Theoretical modelling of electroporation: electrical behavior and molecular transport, in *Electricity and Magnetism in Biology and Medicine* (Blank, M., ed.), San Francisco, pp. 138–140.

25. Weaver, J. C. and Mintzer, R. A. (1981) Decreased bilayer stability due to transmembrane potentials. *Phys. Lett.* **86A,** 57–59.

26. Benz, R., Beckers, F., and Zimmermann, U. (1979) Reversible electrical breakdown of lipid bilayer membranes: a charge-pulse relaxation study. *J. Membrane Biol.* **48,** 181–204.

27. Pastushenko, V. F. and Chizmadzhev, Yu. A. (1982) Stabilization of conducting pores in BLM by electric current. *Gen. Physiol. Biophys.* **1,** 43–52.

28. Sugar, I. P. and Neumann, E. (1984) Stochastic model for electric field-induced membrane pores: electroporation. *Biophys. Chemistry* **19,** 211–225.

29. Weaver, J. C., Harrison, G. I., Bliss, J. G., Mourant, J. R., and Powell, K. T. (1988) Electroporation: high frequency of occurrence of the transient high permeability state in red blood cells and intact yeast. *FEBS Lett.* **229,** 30–34.

30. Tsoneva, I., Tomov, T., Panova, I., and Strahilov, D. (1990) Effective production by electrofusion of hybridomas secreting monodonal antibodies against Hc-antigen of *Salmonella. Bioelectrochem. Bioenerg.* **24,** 41–49.

31. Weaver, J. C. (1993) Electroporation: a dramatic, nonthermal electric field phenomenon, in *Electricity and Magnetism in Biology and Medicine* (Blank, M., ed.), San Francisco, pp. 95–100.

32. Chernomordik, L. V., Sukharev, S. I., Abidor, I. G., and Chizmadzhev, Yu. A. (1982) The study of the BLM reversible electrical breakdown mechanism in the presence of UO_2^{2+}. *Bioelectrochem. Bioenerg.* **9,** 149–155.

33. Neumann, E. and Rosenheck, K. (1972) Permeability changes induced by electric impulses in vesicular membranes. *J. Membrane Biol.* **10,** 279–290.

34. Kinosita, K. Jr. and Tsong, T. Y. (1978) Survival of sucrose-loaded erythrocytes in circulation. *Nature* **272,** 258–260.

35. Klenchin, V. A., Sukharev, S. I., Serov, S. M., Chernomordik, L. V., and Chizmadzhev, Yu. A. (1991) Electrically induced DNA uptake by cells is a fast process involving DNA electrophoresis. *Biophys. J.* **60,** 804–811.

36. Sukharev, S. I., Klenchin, V. A., Serov, S. M., Chernomordik, L. V., and Chizmadzhev, Y. A. (1992) Electroporation and electrophoretic DNA transfer into cells. *Biophys. J.* **63,** 1320–1327.

37. Prausnitz, M. R., Lau, B. S., Milano, C. D., Conner, S., Langer, R., and Weaver, J. C. (1993) A quantitative study of electroporation showing a plateau in net molecular transport. *Biophys. J.* **65,** 414–422.

38. Prausnitz, M. R., Milano, C. D., Gimm, J. A., Langer, R., and Weaver, J. C. (1994) Quantitative study of molecular transport due to electroporation: uptake of bovine serum albumin by human red blood cell ghosts. *Biophys. J.* **66,** 1522–1530.

39. Gift, E. A. and Weaver, J. C. (1995) Observation of extremely heterogeneous electroporative uptake which changes with electric field pulse amplitude in *Saccharomyces cerevisiae. Biochim. Biophys. Acta* **1234(1),** 52–62.

40. Hui, L., Gift, E. A., and Weaver, J. C. Uptake of Bovine Serum Albumin by Yeast due to Electroporation: Existence of a Plateau as Pulse Amplitude is Increased (in preparation).

41. Lillie (1958) Glass, in *Handbook of Physics* (Condon, E. U. and Odishaw, H., eds.), McGraw-Hill, New York, pp. 8–83, 8–107.

42. Neumann, E., Sprafke, A., Boldt, E., and Wolf, H. (1992) Biophysical digression on membrane electroporation, in *Guide to Electroporation and Electrofusion* (Chang, D. C., Chassy, B. M., Saunders, J. A., and Sowers, A. E., eds.), Academic.

43. Lee, R. C., River, L. P., Pan, F.-S., Ji, L., and Wollmann, R. L. (1992) Surfactant induced sealing of electropermeabilized skeletal muscle membranes *in vivo. Proc. Natl. Acad. Sci. USA* **89,** 4524–4528.

44. Gift, E. A. and Weaver, J. C. (1993) Cell survival following electroporation: quantitative assessment using large numbers of microcolonies, in *Electricity and Magnetism in Biology and Medicine* (Blank, M., ed.), San Francisco, pp. 147–150.

45. Weaver, J. C., Bliss, J. G., Powell, K. T., Harrison, G. I., and Williams, G. B. (1991) Rapid clonal growth measurements at the single-cell level: gel microdroplets and flow cytometry. *Bio/Technology* **9,** 873–877.

46. Weaver, J. C., Bliss, J. G., Harrison, G. I., Powell, K. T., and Williams, G. B. (1991) Microdrop technology: a general method for separating cells by function and composition. *Methods* **2,** 234–247.

47. Weaver, J. C. (1994) Molecular basis for cell membrane electroporation. *Ann. NY Acad. Sci.* **720,** 141–152.

48. Okino, M. and Mohri, H. (1987) Effects of a high-voltage electrical impulse and an anticancer drug on *in vivo* growing tumors. *Jpn. J. Cancer Res.* **78,** 1319–1321.

49. Mir, L. M., Orlowski, S., Belehradek, J., Jr., and Paoletti, C. (1991) *In vivo* potentiation of the bleomycin cytotoxicity by local electric pulses. *Eur. J. Cancer* **27,** 68–72.

50. Dev, S. B. and Hofmann, G. A. (1994) Electrochemotherapy—a novel method of cancer treatment. *Cancer Treatment Rev.* **20,** 105–115.

51. Prausnitz, M. R., Bose, V. G., Langer, R. S., and Weaver, J. C. (1992) Transdermal drug delivery by electroporation. Abstract, Proc. Intern. Symp. Control. Rel. Bioact. Mater. 19, Controlled Release Society, July 26–29, Orlando, FL, pp. 232,233.

52. Prausnitz, M. R., Bose, V. G., Langer, R., and Weaver, J. C. (1993) Electroporation of mammalian skin: a mechanism to enhance transdermal drug delivery. *Proc. Natl. Acad. Sci. USA* **90,** 10,504–10,508.

53. Titomirov, A. V., Sukharev, S., and Kistoanova, E. (1991) In vivo electroporation and stable transformation of skin cells of newborn mice by plasmid DNA. *Biochim. Biophys. Acta* **1088,** 131–134.

54. Sukharev, S. I., Titomirov, A V., and Klenchin, V. A. (1994) Electrically-induced DNA transfer into cells. Electrotransfection in vivo, in *Gene Therapeutics* (Wolff, J. A., ed.), Birkhäuser, Boston, pp. 210–232.

55. Gaylor, D. C., Prakah-Asante, K., and Lee, R. C. (1988) Significance of cell size and tissue structure in electrical Trauma. *J. Theor. Biol.* **133,** 223–237.

56. Bhatt, D. L., Gaylor, D. C., and Lee, R. C. (1990) Rhabdomyolysis due to pulsed electric fields. *Plast. Reconstr. Surg.* **86,** 1–11.

57. Hughes, K. and Crawford, N. (1989) Reversible electropermeabilisation of human and rat blood platelets: evaluation of morphological and functional integrity "in vitro" and "in vivo." *Biochim. Biophys. Acta* **981,** 277–287.

58. Mouneimne, Y., Tosi, P.-F., Barhoumi, R., and Nicolau, C. (1991) *Biochim. Biophys. Acta* **1066,** 83–89.

59. Zeira, M., Tosi, P.-F., Mouneimne, Y., Lazarte, J., Sneed, L., Volsky, D. J., and Nicolau, C. (1991) *Proc. Natl. Acad. Sci. USA* **88,** 4409–4413.
60. Belehradek, M., Domenge, C., Orlowski, S., Belehradek, J., Jr., and Mir, L. M. (1993) *Cancer* **72,** 3694–3700.
61. Riviele, J. E., Monterio-Riviere, N. A., Rogers, R. A., Bommannan, D., Tamada, J. A., and Potts, R. O. Pulsatile Transdermal Delivery of LHRH Using Electroporation: Drug Delivery and Skin Toxicology (submitted).
62. Potts, R. O. and Francoeur, M. L. (1990) Lipid biophysics of water loss through the skin. *Proc. Natl. Acad. Sci. USA* **87,** 3871–3873.
63. Bach, D. and Miller, I. R. (1980) Glyceryl monooleate black lipid membranes obtained from squalene solutions. *Biophys. J.* **29,** 183–188.
64. Sugar, I. P. (1981) The effects of external fields on the structure of lipid bilayers. *J. Physiol. Paris* **77,** 1035–1042.

CHAPTER 2

Effects of Pulse Length and Strength on Electroporation Efficiency

Sek Wen Hui

1. Introduction

Electroporation is now a standard method of transfection and cell load-ing. There is a variety of commercial electroporation equipment, and many published and manufacturer-supplied protocols. Many of these pro-tocols are results of trial and error. These empirical protocols are valu-able guides for successful applications of electroporation.

Because experimental conditions vary case by case, new and modified protocols are constantly needed to optimize the transfection yield. Devel-oping new protocols for new cases by trial and error in each laboratory is wasteful in terms of time and energy. This chapter is intended to present a guide, based on known theories of electroporation, to help users modify existing protocols and develop new protocols for new applications. Suc-ceeding in doing so would take some guess work out of experimental trials in tailoring protocols for individual case needs.

2. Theoretical Guide
2.1. General Considerations

Although the detailed molecular mechanism of electroporation is still not completely understood, recent fundamental studies give us a general concept of the events happening during the electroporation process. We know that membrane permeation is the result of the electric breakdown of the lipid bilayer when the induced transmembrane potential exceeds the breakdown potential of the bilayer. Pores in membranes are formed

From: *Methods in Molecular Biology, Vol. 55: Plant Cell Electroporation and Electrofusion Protocols* Edited by: J. A. Nickoloff Humana Press Inc., Totowa, NJ

and maintained by the electric pulse field. In contrast to lipid bilayers that reseal immediately after the breakdown, the permeated state of cell membranes may last tens of minutes after the termination of the pulse field. During this time, molecules may enter the cell through a number of pathways, including electrophoresis (of charged molecules, such as DNA), electro-osmotic and colloid-osmotic flow, as well as diffusion. Therefore, several physical factors may affect the electro-transfection efficiency:

1. The transmembrane potential created by the imposing pulse electric field.
2. The extent of membrane permeation (number and size of pores or affected areas).
3. The duration of the permeated state.
4. The mode and duration of molecular flow.
5. The global and local (surface) concentrations of DNA.
6. The form of DNA.
7. The tolerance of cells to membrane permeation.
8. The heterogeneity of the cell population.

We now examine these factors more quantitatively.

2.2. Creating Electropores

A membrane pore is created when the energy stored in the membrane capacitor exceeds the energy required to keep the membrane intact against pore expansion. The energy E_p of forming a pore of a given radius r in a membrane is determined by the balance between the line tension γ of the pore edge and the surface tension Γ of the membrane.

$$E_p = 2\pi r\gamma - \pi r^2\Gamma \tag{1}$$

This pore energy reaches a maximum at a critical value of pore radius $r_c = \gamma/\Gamma$ *(1,2)*. The line tension of the pore edge depends on the molecular packing of the membrane. Pores with radii $<r_c$ tend to reseal, whereas those with radii $>r_c$ tend to expand if the membrane is under tension.

When subject to an imposed electric field E, which causes the charging of the equivalent membrane capacitor, the energy stored in this membrane capacitor over a precursor area of the pore is:

$$E_e = \pi r^2\varepsilon_0(\varepsilon_w - \varepsilon_m)V^2/2d \tag{2}$$

where ε denotes the dielectric constant and the subscripts *0*, *w*, and *m* refer to free space, water, and membrane, respectively. V is the transmembrane voltage (membrane potential) imposed by the pulse field across the membrane of thickness d. For a given membrane, an electropore of radius r will form if the electric energy E_e given in Eq. (2) is greater than the energy E_p required to form a pore of such size. This energy defines the breakdown voltage V_b, over which the membrane will break down and pores of diameter r will be formed. If $r < r_c$, the electric breakdown is reversible. Otherwise the pore will expand once it is formed. A macroscopic relationship has been described by Zhelev and Needham *(2)*.

In considering the simple case of a spherical cell, the electric conductance of the cell interior is much higher than that of the cell membrane. The imposed membrane potential, or transmembrane voltage V experienced by the cell is:

$$V = 1.5 \, a \, E \cos \theta \, [1 - \exp(-t/\tau)] \tag{3}$$

where E is the pulse electric field, a is the radius of the cell, and θ is the angle between the field direction and the radial vector of the surface point where membrane potential is considered. The charging or relaxation time τ of the membrane is determined by the internal and external conductivities of the cell *(3)*. The highest V is, of course, at the poles along or against the field direction, where θ is 0 and π respectively. Since the breakdown voltage, V_b, of most biomembranes is about 1 V, for a cell 10 μm in diameter, a 0.7 kV/cm pulse is sufficient to produce a breakdown potential at the poles. As the imposed field strength E increases, the area where membrane breakdown potential is experienced extends further from the pole, i.e., the wider the breakdown area. The percentage of breakdown area is given by $(1 - E_b/E)$ where E_b is the pulse field strength needed to produce the membrane breakdown voltage V_b *(4)*.

2.3. DNA Transport

Apart from the extent of membrane permeation, there are several other factors that control the intake of exogenous DNA by the cell. It is believed that the majority of exogenous DNA transported into cells in electroporation is through electrophoresis *(5,6)*. Even if cells remain permeated long after the pulse, as determined by dye penetration, adding DNA immediately after the pulse usually results in a much lower transfection efficiency compared to adding DNA before the pulse. Shielding the charge of DNA by cations also reduces the transfection efficiency. It has been reported that more DNA enters the cells during a second pulse of lower field strength, once the cell is permeated by the first pulse *(7,8)*. Uptake of DNA adsorbed on cell surfaces has also been suggested *(9)*. For a given pore-forming pulse voltage, the transfection efficiency depends more on the total length of a pulse than on the time span when cells remain permeable. Thus, the diffusion of DNA into cells through electropores is not an important contribution in transfection efficiency. Other physical factors, such as the form and concentration of DNA, are also important *(10)*. Because the physical forms of DNA affect their electrophoresis mobility, among other effects, these factors are important with regard to the pulse length and strength.

If the majority of DNA enters the cell during the pulse, the transfection efficiency should be proportional to the time integral of the applied field multiplied by the extent of membrane permeation. For simple rectangular pulses or exponential decay pulses, the time integral is the pulse length (or decay time constant) T, which is usually much greater than the membrane relaxation time τ. Thus, we expect the transfection efficiency to be proportional to:

$$\int E(1 - E_b/E)dt = (E - E_b)T \qquad (4)$$

If multiple rectangular pulses are used, T represents the sum of all pulse periods. If an exponentially decaying pulse from a capacitor-type pulse generator is used, T represents the decay time constant of the pulse. The relationship should hold as long as E is not too much greater than E_b, such that the electropores so created are reversible, the permeated areas stay in the vicinity of cell poles, and the viability of the cell population is not significantly compromised *(4,11)*.

It should be pointed out that many electroporation protocols give electric parameters in terms of voltage and capacitor value (in μF, for instance). These quantities are meaningful only if the sample resistance and interelectrode distance are given as well. For a uniform sample in a cuvet, the pulse field strength E is the applied voltage divided by the interelectrode distance. The pulse decay time $T = RC$ is the capacitance C used in the pulse generator multiplied by the total resistance R of the sample and any circuit elements. The formulae relating voltage and capacitance to the relevant electric parameters E and T are sample- and instrument-dependent. Therefore, voltage and capacitance should not be cited as universal electric parameters. Because sample resistance varies case by case, users should convert the capacitance setting to decay time constant for reporting purposes.

2.4. Sample Homogeneity and Cell Viability

From experience, we know that not every cell in a given sample is transfected, even if the above conditions are satisfied. In fact, only a small percentage of cells are transfected in each run. Within a given sample, cells vary in size and shape, as well as in the dielectric and conductive properties of cellular components, such that the critical field and membrane relaxation time differ from cell to cell. In addition to this physical variation are the more important biological variations in cell cycle, age, and gene expression controls. Therefore, these theoretical considerations should be treated only as a guide. The actual response will depend on the homogeneity of the cell population.

The above analysis applies only to reversible breakdown of cell membranes by the electric pulses, such that membranes reseal in time, and cells recover from the traumatic event of electroporation. In cases where the applied field is high enough to trigger irreversible membrane breakdown, i.e., $E >> E_b$ such that $r > r_c$, electropores do not reseal, and the cell viability is low. As a consequence, cell viability imposes an upper limit on the transfection efficiency.

3. Experimental Evidence

Several experiments using fluorescent molecules as tracers have shown that macromolecule leakage or intake during electroporation is indeed proportional to the quantity $(E - E_b)T$ *(4,12,13)*. Figure 1 shows the uptake of fluorescein isothiocyanate-labeled dextran by mouse C3H/10T$_{1/2}$ cells as a function of ET. Because cells take up dextran spontaneously, the x-axis intercept does not correspond to the value of E_bT. Even though the experimental data represent a wide range of E and T combinations, the linear relationship is obvious, regardless of voltage and time ranges, as long as the reversible breakdown limit is not exceeded *(12)*. A recent measurement of bovine serum albumin uptake by erythrocyte ghosts shows the same trend *(13)*.

Short-term transfection efficiency is sometimes expressed in terms of the percentage of cells transfected or the number of transfected cells per

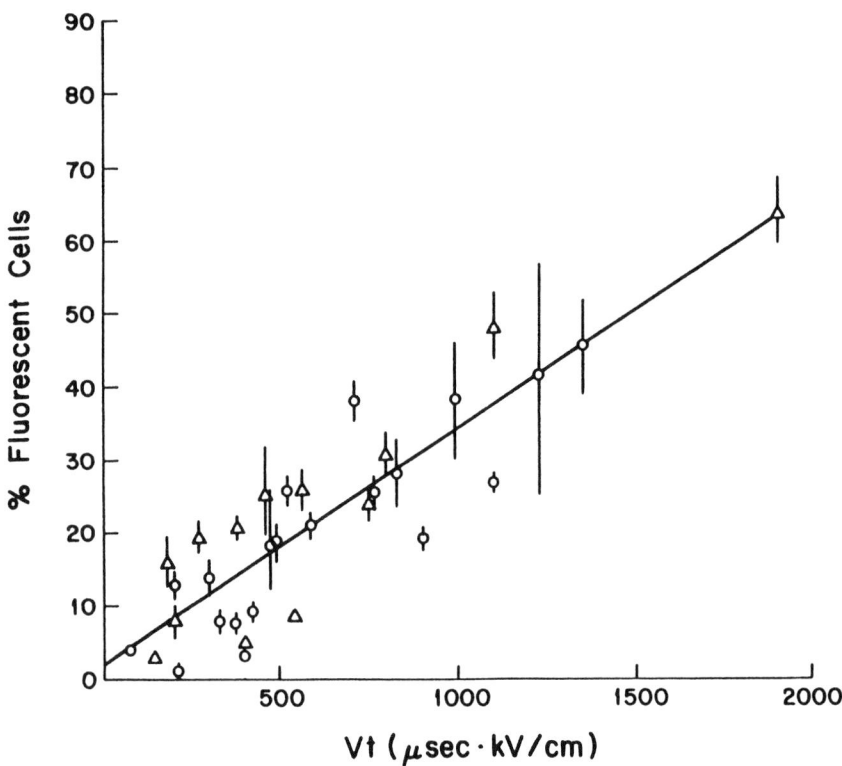

Fig. 1. Macromolecular uptake is proportional to *ET*. The percentage of fluorescent C3H/10T 1/2 cells after electroporation is plotted against the product of pulse length *T* and field strength *E*, for both rectangular (○) and exponential decay (△) pulses, using FITC dextran of mol wt 41 kDa. The straight line represents linear regression of all points *(12)* (courtesy of Eaton Publishing Co.).

given amount of DNA. Both short- and long-term transfection efficiencies are often quoted as the number of clones from an initial population of cells. Whichever transfection efficiency values are measured, these values depend on the percentage of permeated, viable cells, as well as the number of copies of plasmid DNA delivered into those cells.

If the transfection efficiency is proportional to the percentage of permeated cells that receives DNA, then the transfection efficiency would also be a linear function of $(E - E_b)T$. For a given pulse duration, *T*, the transfection efficiency is expected to be proportional to *E*. Similarly, for a given *E*, the transfection efficiency is expected to be proportional to *T*.

The effects of these electric parameters on pBR322 transfection of *E. coli* JM105 were reported by Xie and Tsong *(14)*. Figure 2 shows such relations plotted as log[transfection efficiency*(TE)]* against *E* or log*[T]*. The first plot (Fig. 2A) is not expected to be linear if DNA enters the cell mainly by electrophoresis rather than by diffusion *(4)*, but if log*[TE]* is plotted against log*[E]*, a more linear relationship is found. A linear relationship is also apparent in plots of log*[TE]* against log*[T]* plot (Fig. 2B). Furthermore, when cell viability is affected by irreversible membrane breakdown, the linear relationship yields to the viability limit.

The transfection efficiency of HeLa cells by pRSVgpt plasmid DNA *(15)* was measured as a function of *ET* (*T* is given as the exponential decay half time $\tau_{1/2}$ of pulses generated by a capacitor-type generator). Figure 3 shows that the approximately linear relationship is obeyed by both the short (0.275–0.310 ms) and the long (2.2–4.4 ms) pulse groups. Apparently, even at the highest voltage applied, the cell viability limit had not been reached. The *x*-axis intercept of the least-square-fit line gives $E_b T$ = 0.5 kV ms/cm. This value implies that, for a 0.75-ms pulse, the threshold applied field strength to cause reversible breakdown in some cells is 0.7 kV/cm. This value is approximately equal to that given by Eq. (3), and agrees with most threshold field strength values for electroporation and electrofusion *(16)*.

Human lymphoid cells were transfected by pCP4-fucosidase plasmids. Cells were subjected to three consecutive exponentially decay pulses while in Baker and Knight *(17)* medium. The cells were incubated at 37°C in the same medium for 30 min after the pulses, before transferring back to the normal culture medium. The fucosidase activity was assayed after 48 h. The transfection efficiency given by the enzyme activity of transfected cells is shown in Fig. 4. Data points were taken within the range of field strengths of 0.4–4.0 kV/cm, and durations of 0.14–3.4 ms. Apparently, the viability limit is reached at 1.5 kV ms/cm. Below this limit, the transfection efficiency is approximately linear with *ET*. The threshold breakdown condition is again about $E_b T$ = 0.5 kV ms/cm. A single point (solid square) obtained using 4 kV/cm field strength pulses is exceptionally low in transfection efficiency, perhaps owing to too much cell death caused by irreversible membrane permeabilization.

The effect of pulse strength and duration on the transfection of CHO cells was investigated by Wolf et al. *(6)*. The transfection efficiency of CHO cells in suspension, by pSV2CAT or pBR322-βgal plasmids, was

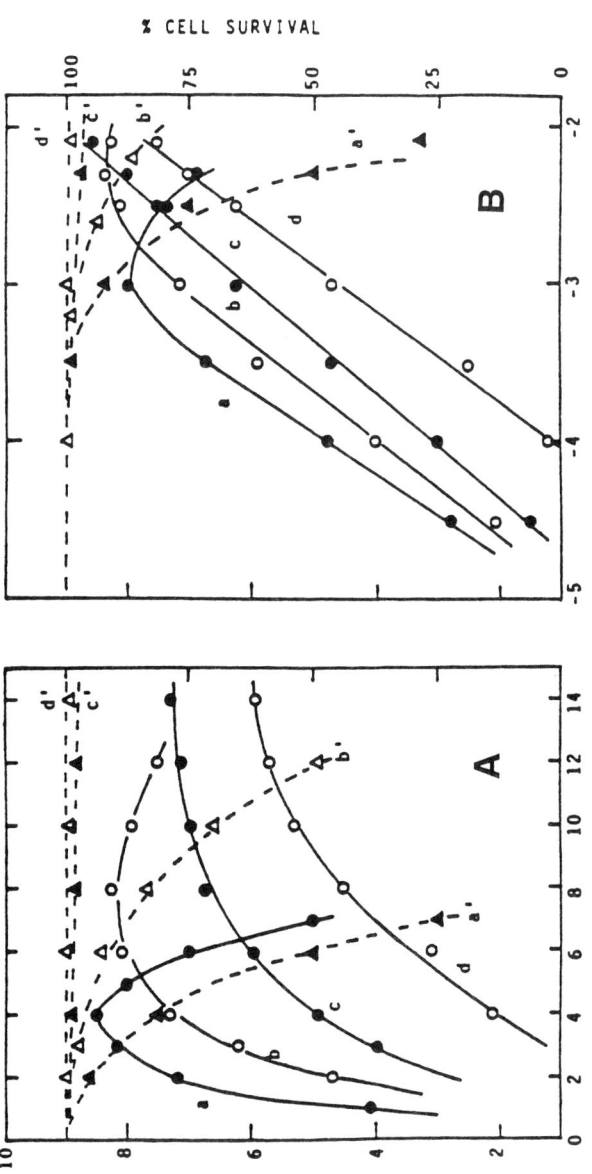

Fig. 2. Effects of electric parameters on pBR322 transfection of *E. coli* JM105. **(A)** Dependence of log[TE] on the pulse field strength. The pulse durations were 5 ms, 1 ms, 200 μs, and 40 μs for curves *a*, *b*, *c*, and *d*, respectively. The corresponding curves (- - -) for percent cell survival are given in curves *a'*, *b'*, *c'*, and *d'*. The left ordinate indicates logarithm of the transfection efficiency (TE) and the right ordinate the percent cell survival. **(B)** Dependence of log[TE] on pulse width. The field strengths were 6, 4, 3, and 2 kV/cm, respectively, for curves *a*, *b*, *c*, and *d*. The corresponding curves (- - -) for percent cell survival are given in curves *a'*, *b'*, *c'*, and *d'*. Experimental conditions are identical to (A) *(14)* (courtesy of the Rockefeller University Press and the Biophysical Society).

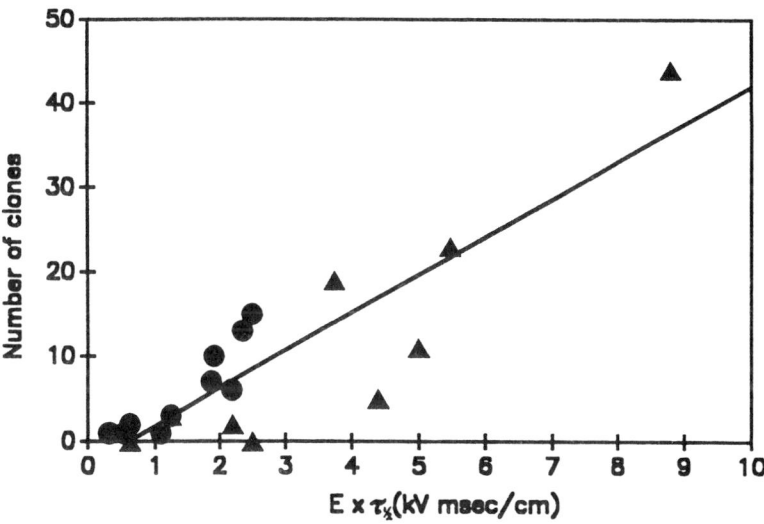

Fig. 3. Effect of electric parameters *ET* on pRSVgpt transfection of HeLa cells. The number of successfully transfected HeLa cells with pRSVgpt plasmid by electroporation is plotted as a function of the product of peak field strength *E* and decay half-time $\tau_{1/2}$ of an applied exponential pulse. Data points for different pulse lengths ($\bullet$ $\tau_{1/2}$ = 0.275–0.310 ms, $\blacktriangle$ $\tau_{1/2}$ = 2.2–4.4 ms) are plotted together *(15)* (courtesy of the Eaton Publishing Co.).

found to be a linear function of either *E* or *T*, when the other parameter was held constant. The minimum field strength E_b for detectable permeability as well as transfection was found to be 0.6 kV/cm. Interestingly, the transfection efficiency decreases with increasing delay between repeating pulses, indicating that DNA is collected on the cell surface and driven through the electropores by electrophoretic force, in agreement with previous experiments using two consecutive pulses to increase transfection yield *(8)*.

4. Conclusion

The existing theory for reversible electric breakdown of cell membranes and the transport of DNA across the plasma membranes through electropores adequately describes a linear relation between transfection efficiency and *(E − E_b)T*. E_b is determined by the electric energy derived from the applied pulses and the energy cost to form an electropore in the

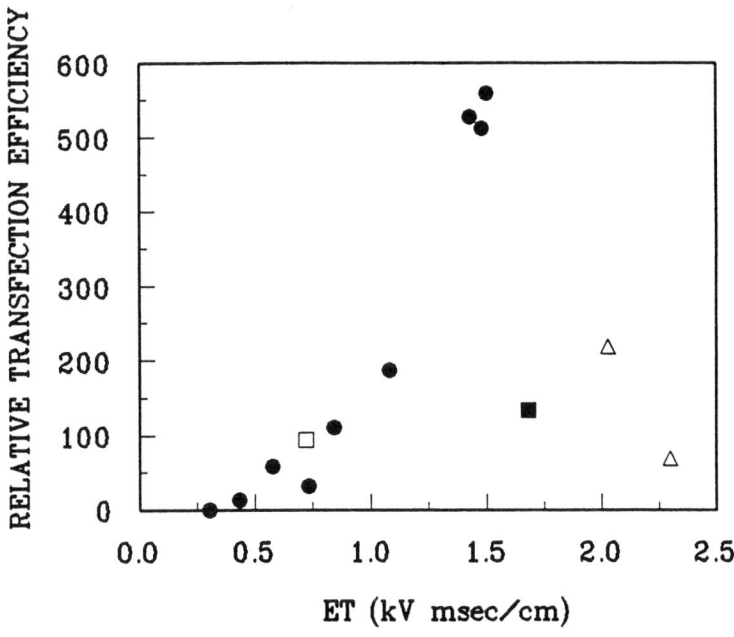

Fig. 4. The relative transfection efficiency of human JTL lymphoid cells transfected by pCEP4-FUC. The activity of α-fusosidase is plotted as a function of the product of the pulse field strength E and the equivalent integeral pulse time T. The samples were subjected to three exponentially decaying pulses of given decay half-times: (□) 0.2 ms, (●) 0.24–0.28 ms, (△) 0.34–0.35 ms, at E = 0.5–2 kV/cm, and (■) 0.14 ms at E = 4 kV/cm.

membrane of cells of a given size. Although the viability limit varies from cell line to cell line, within this limit, the linear relationship between transfection efficiency and $(E - E_b)T$ seems to hold over a wide range of E and T combinations, and is applicable to most cells. Therefore, this relationship serves as a guideline for selection of electric parameters in various applications.

Since the transfection efficiency is proportional to ET, there is a choice of optimizing either E or T. Excessive field strength leads to permeation of a wider area, or forming larger pores that may exceed the reseal limit $r_c = \gamma/T$. As long as pores remain reversible, the required T value may be satisfied by applying longer and multiple pulses. Cell membranes do not reseal as rapidly as lipid bilayers, so that multiple pulses may not form additional pores in the membranes. There are reported advantages of multiple pulses over a single pulse, and AC bursts over rectangular pulses *(18,19)*. A bipolar oscillating field was reported to be more effective than a unipolar oscillating field, since both poles of the cells are permeated *(20)*. Clearly there is merit in using lower applied field strength, and increasing the duration and number of pulses to achieve the required *ET* value.

Protocols for electroporation continue to develop as cases demand. The present knowledge of electroporation mechanisms is capable of guiding rational design of new protocols. Instead of describing a cumulation of successful protocols, this chapter gives the basics to develop one's own protocols. It is hoped that the above analysis will take some guesswork out of applying electrotransfection in new situations.

Acknowledgments

The results from my laboratory cited in this chapter are the effort of R. T. Kubiniec, D. A. Stenger, H. Liang, and N. G. Stoicheva. The work was supported by a grant GM 30969 from the National Institutes of Health.

References

1. Abidor, I. G., Arakelyan, V. B., Chernomordik, L. V., Chizmadzhev, Y. A., Pastushenko, V. F., and Tarsevich, M. R. (1979) Electric breakdown of bilayer membranes. I. The main experimental facts and their qualitative discussion. *Bioelectrochem. Bioenerg.* **6,** 37–52.
2. Zhelev, D. V. and Needham, D. (1993) Tension-stabilized pores in gaint vesicles: determination of pore size and pore line tension. *Biochim. Biophys. Acta* **1147,** 89–104.
3. Stenger, D. A., Kaler, K. V. I. S., and Hui, S. W. (1991) Dipole interaction in electrofusion. Contributions of membrane potential and effective dipole interaction pressures. *Biophys. J.* **59,** 1074–1084.
4. Schwister, K. and Deuticke, B. (1985) Formation and properties of aqueous leaks induced in human erythrocytes by electrical breakdown. *Biochim. Biophys. Acta* **816,** 332–348.
5. Klenchin, V. A., Sukharev, S. I., Serov, S. M., Chernomordik, L. V., and Chidmadzhev, Y. A. (1991) Electrically induced DNA uptake by cells is a fast process involving DNA electrophoresis. *Biophys. J.* **60,** 804–811.
6. Wolf, H., Rols, M. P., Boldt, E., Neumann, E., and Teissie, J. (1994) Control by pulse parameters of electric field-mediated gene transfer in mammalian cells. *Biophys. J.* **66,** 524–531.
7. Andreason, G. L. and Evans, G. A. (1989) Optimization of electroporation for transfection of mammalian cells. *Anal. Biochem.* **180,** 269–275.

8. Sukharev, S. I., Klenchin, V. A., Serov, S. M., Chernomordik, L. V., and Chizmadzhev, Y. A. (1992) Electroporation and electrophoretic DNA transfer into cells. The effect of DNA interaction with electropores. *Biophys. J.* **63,** 1320–1327.

9. Xie, T. D., Sun, L., and Tsong, T. Y. (1990) Study of mechanisms of electric field-induced DNA transfection. I. DNA entry by surface binding and diffusion through membrane pores. *Biophys. J.* **58,** 13–19.

10. Nickoloff, J. A. and Reynolds, R. J. (1992) Electroporation-mediated gene transfer efficiency is reduced by linear plasmid carrier DNAs. *Anal. Biochem.* **205(2),** 237–243.

11. Rols, M. P. and Teissie, J. (1990) Electropermeabilization of mammalian cells. Quantitative analysis of the phenomenon. *Biophys. J.* **58(5),** 1089–1098.

12. Liang, H., Purucker, W. J., Stenger, D. A., Kubiniec, R. T., and Hui, S. W. (1988) Uptake of fluoroscence-labeled dextrans by 10T 1/2 fibroblasts following permeation by rectangular and exponential-decay electric field pulses. *BioTechniques* **6,** 550–558.

13. Prausnitz, M. R., Milano, C. D., Gimm, J. A., Langer, R., and Weaver, J. C. (1994) Quantitative study of molecular transport due to electroporation: uptake of bovine serum albumin by erythrocyte ghosts. *Biophys. J.* **66,** 1522–1530.

14. Xie, T. D. and Tsong, T. Y. (1992) Study of mechanisms of electric field-induced DNA transfection III. Electric parameters and other conditions for effective transfection. *Biophys. J.* **63,** 28–34.

15. Kubiniec, R. T., Liang, H., and Hui, S. W. (1990) Effects of pulse length and pulse strength on transfection by electroporation. *BioTechniques* **8,** 1–3.

16. Chang, D. C., Chassy, B. M., Saunders, J. A., and Sowers, A. E. (1992) *Guide for Electroporation and Electrofusion.* Academic, San Diego.

17. Baker, P. F. and Knight, D. E. (1983) High-voltage techniques for gaining access to the interior of cells: application to the study of exocytosis and membrane turnover. *Methods Enzymol.* **98,** 28–37.

18. Xie, T. D. and Tsong, T. Y. (1990) Study of mechanisms of electric field-induced DNA transfection. II. Transfection by low amplitude low frequency alternating electric fields. *Biophys. J.* **58,** 897–903.

19. Chang, D. C., Gao, P. Q., and Maxwell, B. L. (1991) High efficiency gene transfection by electroporation using a radio-frequency electric field. *Biochim. Biophys. Acta* **1992,** 153–160.

20. Tekle, E., Austumian, R. D., and Chock, P. B. (1991) Electroporation by using bipolar oscillating electric field: an improved method for DNA transfection of NIH 3T3 cells. *Proc. Natl. Acad. Sci. USA* **88,** 4230–4234.

CHAPTER 3

Instrumentation

Gunter A. Hofmann

1. Introduction

The techniques of electroporation and electrofusion require that cells be subjected to brief pulses of electric fields of the appropriate amplitude, duration, and wave form. In this chapter, the term electro cell manipulation (ECM) shall describe both techniques. ECM is a quite universal technique that can be applied to eggs, sperm, platelets, mammalian cells, plant protoplasts, plant pollen, liposomes, bacteria, fungi, and yeast—generally to any vesicle surrounded by a membrane. The term "cells" will be used representatively for any of the vesicles to be manipulated unless specific requirements dictate otherwise.

Electroporation is characterized by the presence of one membrane in proximity to molecules that are to be released or incorporated. One or several pulses of the appropriate field strength, pulse length, and wave shape will initiate this process.

Electrofusion is characterized by two membranes in close contact that can be joined by the application of a pulsed electric field. The close contact can be achieved by mechanical means (centrifuge), chemical means (PEG), biochemical means (avidin-biotin *[1]*), or by electrical means (dielectrophoresis *[2]*). Only the electric method is discussed as it relates to ECM instrumentation.

The intent of this chapter is to provide the researcher with a basic understanding of the hardware components and electrical parameters of ECM systems to allow intelligent, economical choices about the best instrumentation for a specific application and to understand its limita-

From: *Methods in Molecular Biology, Vol. 55: Plant Cell Electroporation and Electrofusion Protocols* Edited by: J. A. Nickoloff Humana Press Inc., Totowa, NJ

tions. Commercial instruments have been available for more than 10 years; the commercial ECM technology has matured and become more costeffective. Rarely is it economical to build one's own instrument. Although articles occasionally appear on how to build an instrument for a few hundred dollars, the plans are generally of poor design, and the cost estimates often do not take into account the researcher's time for electronic development. Furthermore, today's commercial instruments often incorporate measuring circuits for important parameters, which are difficult to develop. The difficulty is that in one housing, there are voltages of many kilovolts and currents of hundreds of Amperes (A) flowing next to low signal/control voltages, typically between 5 and 20 V. Sophisticated design is needed to prevent crosstalk or electromagnetic interference between these different circuits. Thus, it is usually not costeffective to build an instrument unless specific parameters are needed that are not available commercially. Another very important issue is safety. Voltages and currents generated in efficient ECM generators are large enough to induce cardiac arrest. Generators need to be constructed to be safe and foolproof against accidental wrong settings. They must also deliver the pulse to the chamber in such a way that the operator will not, under any circumstances, come in contact with parts carrying high voltage.

A database of over 2500 publications in the field of electroporation and electrofusion is maintained and updated continuously by BTX (San Diego, CA) as a service to the research community. Any researcher may inquire about the BTX Electronic Genetics® Database and request a database search.

2. Components of an ECM System and Important Parameters

Generally, ECM systems consist of a generator providing the electric signals and a chamber in which the cells are subject to the electric fields created by the voltage pulse from the generator. A third optional component is a monitoring system, either built into the generator or connected in line between the generator and the chamber, which measures the electrical parameters as the pulse passes through the system. Each component is discussed in Sections 4.–6. In this section, we discuss the relationship between the electrical parameters, which the ECM system provides, and the parameters that the cells experience.

The biophysical process of electropermeabilization is caused by the electrical environment of the cell in a medium. The main parameter, which describes this environment, is the electric field strength E, measured in V/cm. Though the presence of the cell itself modifies the field in close proximity, knowledge of the average field strength at the location of the cells is sufficient for the purpose of ECM experiments. The electric field is generally created by the application of a potential difference (voltage) between metallic electrodes immersed in the medium containing the cells. For the simple electrode geometry of parallel plates located at a distance d (cm), the electric field is calculated from the applied voltage V as:

$$E = V/d \text{ (V/cm)} \tag{1}$$

Practical values of E used in ECM range from a few hundred V/cm for mammalian cells to many kV/cm for bacteria.

The electric field in the medium gives rise to currents depending on the medium specific resistivity r, which is measured in $\Omega \cdot$ cm. The specific resistivity ranges from a low of about 100 $\Omega \cdot$ cm for saline solutions to many k$\Omega \cdot$ cm for nonionic solutions, such as mannitol. The resulting current density j is:

$$j = E/r \text{ (A/cm}^2) \tag{2}$$

The current produced results in heating of the medium. Saline solutions with a low value of r experience severe heating effects as compared to nonionic solutions for the same electric field and pulse length.

The temperature rise ΔT (°C) can influence the permeabilization mechanism, or lead to excessive heating and evaporation of the medium. It can be calculated for different pulse wave shapes:

Square pulse: $\Delta T = E^2 t/4.2 \ r$, where t is the pulse length in s.
Exponential pulse: $\Delta T = E^2 \tau/8.4 \ r$, where τ is the 1/e time constant (s) (*see* Section 4.2.1.).

Having defined the parameters at the location of the cells, we can relate them to the electrical parameters at the chamber electrodes:

Electrode voltage: $V = E \cdot d$ (V) (plane parallel electrodes)
Chamber current: $J = j \cdot F$ (A), where F is the electrode area, cm^2
Chamber resistance: $R = f(r) \ \Omega$, where $f(r)$ is a function of the chamber geometry.
For plane parallel electrodes, $R = r \ d/F$.

The voltage at the chamber electrodes is not necessarily equal to the voltage that is indicated or even measured in the generator. This relationship is discussed in Section 4.1. The range of the electric field for optimum yield is quite narrow. Deviation of 5–10% from the optimum value can lead to a drop of an order of magnitude in yield. The pulse length is a less sensitive parameter. It is desirable to assure that the optimum electric field is established in the chamber.

3. Volume Requirements

Small volumes of 100 µL to a few milliliters can be treated in a batch mode: fill the chamber, electroporate, or fuse, and empty the chamber. Larger volumes (many milliliters to 1 L) require chambers that might not be available and a high output power level, which generators typically cannot deliver. A good solution to this problem is the use of a flow-through system in which the generator periodically pulses in synchronism with a pump, so that every volume element of cell/transformant mixture is exposed to the desired electric fields and number of pulses as it passes through the chamber. This method requires flowthrough chambers and generators that can pulse automatically, either at a fixed or adjustable repetition rate. Such generators and chambers are available (*see* Tables 1 and 6). For fusion, a continuous flow is not desirable, because fused cells need to be undisturbed for a period of time to round off and complete the fusion process. In this case, a pulsating (stop and go) flowthrough system would be appropriate.

4. Generators

The relationship between the electrical parameters in a generator and the parameters actually delivered to the chambers is important because substantial differences can exist. Following this discussion, different types of generators are described. Table 1 presents a survey of commercially available generator types.

4.1. Actual Voltage Delivered to the Chamber

The momentary power the generators are required to deliver to chambers can far exceed the electrical power available from laboratory outlets. To overcome this limitation, electrical energy is stored in capacitors by charging them slowly at low power to a preset voltage and then discharging them at high power level into the chamber. The voltage V_0 to which the capacitors will be charged can be set and is typically indicated

Table 1
Survey of Electroporation and Electrofusion Generators

Manufacturer	No PS	With PS one pulse length	With PS multiple pulse lengths	Square wave	Electro cell fusion	Stand-alone monitor
IBI (New Haven, CT)			X			
Invitrogen (San Diego, CA)	X					
Bio-Rad (Richmond, CA)		X	X			
BRL (Grand Island, NY)		X	X			
BTX	X	X^a	X^a	X^a	X	X

Abbreviation: PS, power supply.
[a]R, optional version available with repetitive pulsing for flowthrough applications.

at the front panel of the generators. The actual voltage delivered to the chamber can be substantially lower than what is normally assumed to be the generator output voltage. This effect is caused by the internal resistance of the generator (typically around 1 Ω), which absorbs part of the charging voltage during discharge and is more pronounced in larger chambers (several milliliters) and low resistivity medium. Some generators are also designed with a relatively high internal resistance, which is undesirable, to protect the output switch against high currents. These generators can exhibit a drastic drop in actual voltage delivered to the chamber under certain circumstances. If the internal resistance R_i and the chamber resistance R_c are known, the actual voltage V on the chamber can be calculated as:

$$V = V_0 \cdot R_c/(R_c + R_i) \tag{3}$$

4.2. Generators for Electroporation

The two types of generators commonly encountered differ by the wave shape of their output: exponential decay wave form or square pulses. Though both can in principle be used for electroporation, it appears that bacteria are transformed more efficiently by exponential wave forms (with some exceptions [3]), whereas some mammalian cell types (4) and plant protoplasts (5) show generally superior transformation results with square waves.

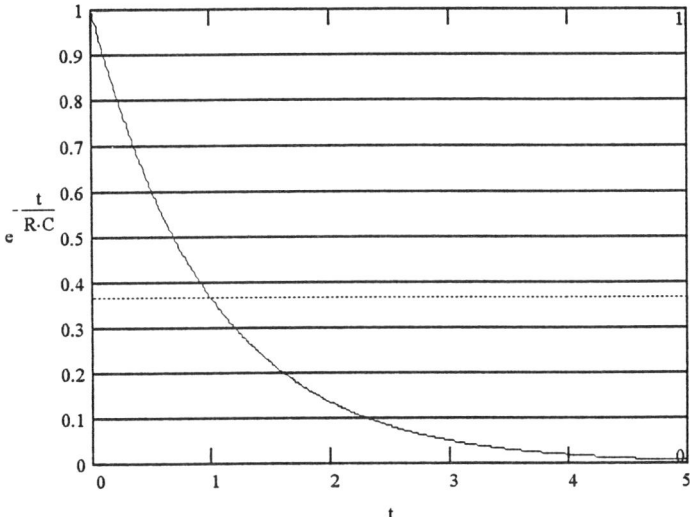

Fig. 1. Exponential decay wave form, representative of the complete discharge of a capacitor into a resistor.

4.2.1. Exponential Wave Form Generators

The voltage of a capacitor C (capacity measured in Farad or, more conveniently, in microfarad) discharging into a resistor R (Ω) follows an exponential decay law (Fig. 1):

$$V = V_o \cdot exp\,(-t/RC) \tag{4}$$

The pulse length of such a discharge wave form is commonly characterized by the "$1/e$ time constant." This is the time required for the initial voltage to decay to $1/e \approx 1/3$ of the initial value ($e = 2.718\ldots$ is the basis of natural logarithms). This time constant can be conveniently calculated from the product of R and C, where C is the storage capacitor in the generator and R is the total resistance into which the capacitor discharges, which can have several components. Figure 2 shows a general circuit diagram of an exponential decay generator.

The power supply slowly charges the capacitor to the desired voltage and does not play a role during the discharge. The internal resistance R_i of the capacitor is on the order of 0.5–1 Ω for electrolytic capacitors and, in normal operation, is much smaller than any other resistance in the circuit and can therefore be neglected. The resistor R_L is installed in some instruments to limit the current in the circuit, especially in case of an arc in the chamber, which would result in high currents because the chamber

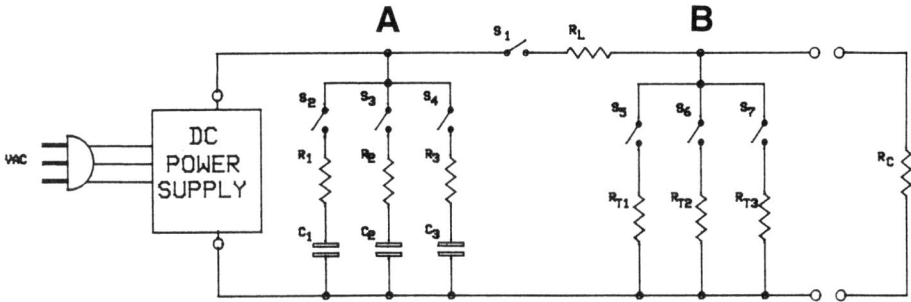

Fig. 2. General circuit diagram of an exponential decay wave form generator. $C_{1,2,3}$ are the energy storage capacitors, which have an internal resistance $R_{1,2,3}$. They can be added to the circuit by switches $S_{2,3,4}$ in order to vary the total capacitance. Closing switch S_1 allows the charged capacitors to discharge to the output and into the chamber, represented by the resistor R_c. R_L is a discharge current-limiting resistor, which is needed in some designs. R_{T1}, R_{T2}, and R_{T3} are timing resistors, which can be added to the circuit by switches $S_{5,6,7}$. If the output voltage is measured at **(A)**, instead of **(B)**, incorrect readings of the actual voltage on the chamber will result.

resistance drops to very low values during an arc. The size of this resistor is determined by the maximum current capability of the switch. As a result of the presence of R_L, the voltage at the chamber is reduced by the voltage drop across R_L, which can be substantial. Furthermore, some instruments measure the peak discharge voltage at the point A instead of directly across the chamber at point B, resulting in incorrect readings. Use of instruments that do not have a built-in current-limiting resistor provides advantages. One needs to be aware that without a current-limiting resistor, arcs appear more violent because of higher current flow. However, if the instrument and chamber stand are designed correctly, this should be of no consequence. It should be noted here that arcing in the chamber occurs mostly at high field strengths (above 10 kV/cm) and is a statistical effect.

The resistance R_t is a timing resistor that, typically, can be selected to adjust the pulse length. Maintaining a low value, relative to the chamber resistance R_c, serves the function of determining pulse length. Often, the size of the capacitance can be changed by connecting one or more capacitors in parallel. Since the time constant is determined by the product of resistance and capacitance, either variable can be used to adjust it. Keeping the resistance as low as possible, well below the chamber resis-

Table 2
Comparison of Electroporation Generators
with Built-in Power Supply and Multiple Pulse Length

Manufacturer of electroporation system	IBI	Bio-Rad	BRL	BTX
Model number	geneZAPPER™ 450/2500	Gene Pulser®	Cell-Porator™	ECM® 600
# of Unit components required	2	3	2	1
Voltage range	50–2500	50–2500	0–2500	50–2500
# of Pulse length settings	34	8	8	126
Maximum field strength (kV/cm)	12.5	25	16.6	40
Maximum current, A	130	120	40	>1000
Monitoring	Partial	Partial	Partial	Full
Actual voltage	No	No	Yes	Yes
Actual pulse length	Yes	Yes	No	Yes
Safety design	Operator shock proof	Safety interlock	Safety interlock	Operator shock, arc, and short circuit proof
Data base	No	No	No	Yes
Warranty	1 yr	1 yr	1 yr	2 yr

tance, is generally desirable. Sometimes the chamber resistance is too low for the timing resistors to be effective. In this case, the chamber resistance itself will determine the pulse length, which then can be adjusted only by varying the capacitance.

For the characterization of the pulse into the chamber, only two parameters need to be known: the peak voltage and the $1/e$ pulse length. It is convenient to use a generator with a built-in measuring circuit that measures the pulse parameters at the output of the instrument (Point B in Fig. 2). Table 2 shows a comparison of the main features of commercially available exponential discharge generators with built-in power supply, multiple pulse length capability, and at least some monitoring.

If only a limited number of applications are planned, such as *E. coli* transformation, a generator with a fixed pulse length will be sufficient. This simplifies the generator design and reduces costs. To reduce costs

Table 3
Exponential Decay Generator Options and Costs

Fixed time constant *t*	Fixed *t*	Variable *t*
No power supply (PS) ~$1000	With PS $1000–2000	With PS and monitoring $4000–5000

even further, it is also possible to use an external electrophoresis power supply and eliminate a built-in supply. Table 3 shows generators available with increasing flexibility and cost options. The maximum voltage is typically 2500 V. The time constant for fixed pulse length is typically 5 ms.

4.2.2. Square-Wave Generators

Square-wave pulses appear to have advantages for certain applications, such as transfection of mammalian cell lines and plant protoplasts, though no generalization can be made. Each cell line needs to be individually investigated to determine whether use of square-wave pulses would be advantageous. In general, square waves do not appear to result in higher transformation yields for bacteria, although there are some protocols that give good results (*6*; Xing Xin, Texas Heart Institute, personal communication). Square waves are used almost exclusively for in vivo applications of electroporation, such as electrochemotherapy, where drugs are electroporated into tumor cells. These generators are more difficult to build because the square-wave pulse is produced by a partial discharge of a large capacitor, which requires the interruption of high currents against high voltages. In the past, their costs were higher than exponential discharge generators, and the range of parameters was more limited. However, recent advances in solid-state switching technology have lowered costs. A square-wave generator is now available that can deliver up to 3000 V into a 20-Ω load at costs comparable to exponential discharge units (*see* Table 1).

4.3. Generators for Electrofusion

If nonelectrical means of cell–cell contact are used, any electroporation generator can also be used for electrofusion. If it is desirable to induce cell–cell contact by dielectrophoresis, the generator needs to produce an alternating wave form (ac) over a longer period of time, typically seconds, before the fusion pulse is applied.

The optimal frequency appears to be around 1 MHz (*7*). Above and below this frequency, the viability of mammalian cells, at least, appears

Table 4
Mouse Egg Fusion with Different Wave Forms and Chambers

AC wave form and chamber type	# of Eggs	% Fusion	% Developed	Stability of development
Nonsinusoidal, wire chamber	20	100	75	Stable
Sinusoidal, rectangular bar chamber	24	87.5	52.4	Less stable

to suffer. Nonionic fusion media are desirable to reduce the generation of heat and turbulence. A pure sinusoidal wave form is not necessary and, possibly, not even advantageous. It is, however, important that there is no net dc component in the wave form. Higher harmonics in the wave form appear to produce better fusion results. Table 4 compares results obtained with different wave forms and chambers *(8)*.

Commercial fusion generators are available (Table 1) that allow the sequential application of ac wave forms and fusion pulses, which are generally of the square-wave type.

4.4. Generators with Other Wave Forms

Researchers have experimented with wave forms other than exponential and square. It is apparent that for some applications, special wave forms have certain advantages. Bursts of radio frequency electric fields (a few 100 kHz) appear to be more benign to cells and might be advantageous when fusing cells of widely different sizes *(9,10)*. However, such generators are not presently available commercially, and are difficult and expensive to build with high-power levels.

5. Chambers

There are many choices in chambers for ECM. In general, chambers need to create the required field strength from the voltage delivered to the electrodes by the generator; they need to contain the appropriate volume, and need to be sterilized or sterilizable, easily filled, emptied, and if reused, easily cleaned. Table 5 gives the main trade-off parameters in the selection of chambers. The following describes only the more frequently used chambers in the field.

Table 5
Chamber Trade-off Parameters

Small volume	Large volume
Disposable	Reusable
Homogeneous field	Inhomogeneous field
Visualization of cells	Cells obscured
Batch process	Flowthrough
Aluminum electrodes	Stainless steel or noble material
Presterilized	Sterilizable
Small gap	Large gap

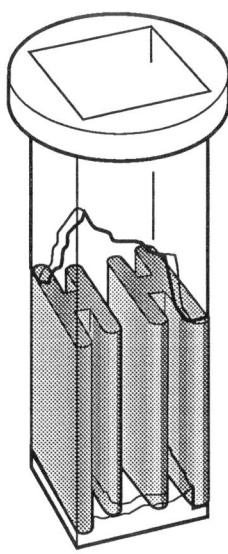

Fig. 3. Disposable electroporation cuvet with molded-in aluminum electrodes.

5.1. Small-Volume Chambers

Disposable, presterilized cuvets with molded-in aluminum electrodes (Fig. 3) are most frequently used for electroporation. They are available with different gap sizes, typically 1 mm (for bacteria), and 2 and 4 mm (for mammalian cells and plant protoplasts). The electric field in these cuvets is quite homogeneous. Some workers clean and reuse cuvets to reduce costs.

Reusable, parallel plate electrode assemblies (Fig. 4) that fit into spectrophotometer cuvets are available. They are also available in different gap sizes.

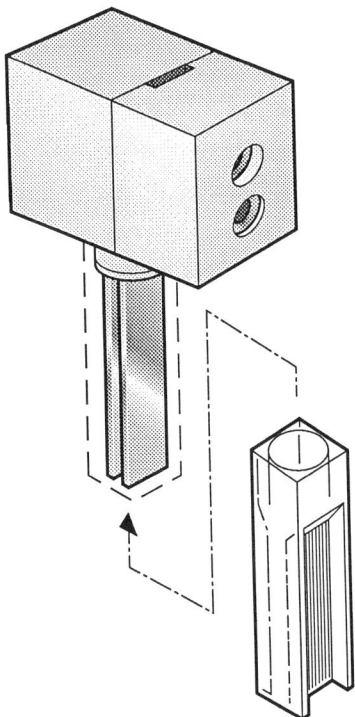

Fig. 4. Reusable parallel plate electrode assembly to fit into spectrophoto-meter cuvets.

Note that chemical cleaning or even autoclaving might not remove cell debris, transformant, and medium breakdown products that might have been deposited onto the electrodes during a pulse. Only a good mechanical cleaning will remove the debris.

Electrodes on microslides are used to visualize the fusion process under a microscope. Parallel wires (Fig. 5), separated by 1 mm or less, produce divergent fields that favor dielectrophoretic pearl chain formations of cells. For small gaps (<1 mm), a meander-type electrode configuration (Fig. 6) allows visualization of the fusion process. Electrodes with square bars (Fig. 7), which provide a more homogeneous electric field, can also be mounted on microslides for visualization of embryo manipulation.

5.2. Large-Volume Chambers

Intermediate-size chambers with a volume of a few milliliters can be built with parallel bars (Fig. 8). The electrodes can be flat to create

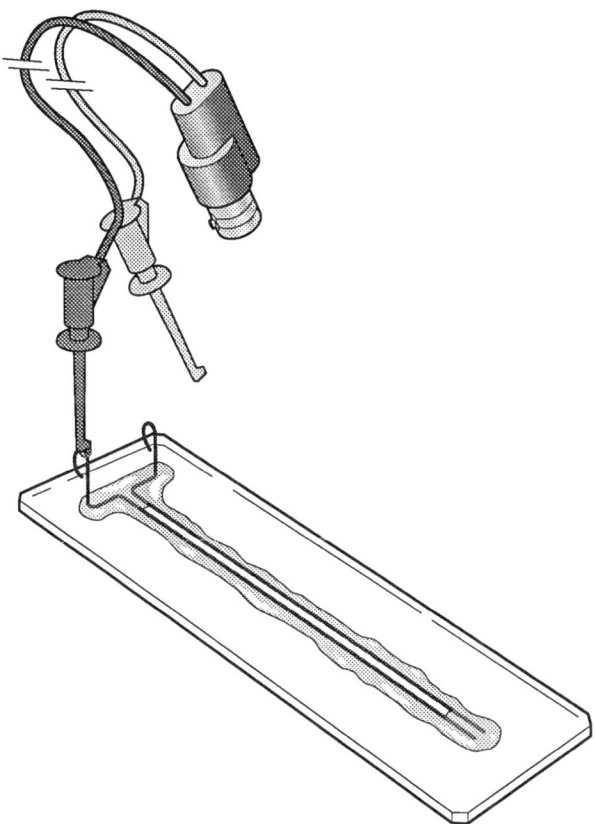

Fig. 5. Parallel wire electrodes mounted on a microslide for visual observation of the electrofusion process. These electrodes create an inhomogeneous electric field, which is preferable for dielectrophoresis.

homogeneous fields or they can have grooves to create divergent fields for fusion. A convenient implementation of a large-volume chamber with a volume up to 50 mL is an array of parallel plate electrodes fitted into a plastic Petri dish (Fig. 9). The gap between the electrodes can be 2 mm for mammalian cells or 10 mm for embryo and fish egg electroporation. Generally, such large volumes need a high resistivity medium because the chamber resistance with saline solution, such as PBS, would be very low. Partial filling of the chamber will reduce the resistance proportionally. As an example, 10 mL of PBS in a 10-cm diameter Petri dish with 2-mm spaced electrodes resulted in a resistance of 0.4 Ω. Some generators can generate sufficient voltage to transform mammalian cells even

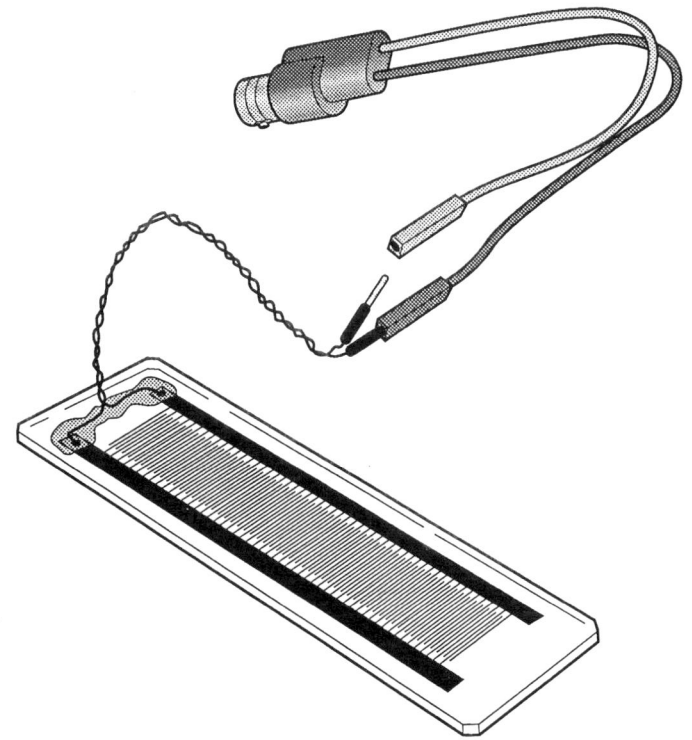

Fig. 6. Meander-type chamber for visual observation of fusion.

with PBS. The parallel plate electrode configuration in a Petri dish is also very useful for electroporation of adherent cells, if the electrodes are situated so they touch the Petri dish bottom. Instead of parallel plates, an array of concentric electrodes can also be used to create a large-volume ECM chamber in a Petri dish *(11)*.

5.3. Small-Volume Flowthrough Chambers for the ECM of Large Volumes

If it is required to transform large volumes (above 50 mL), it is economical to pulse the generator repetitively in synchrony with a pump that pushes the medium with the cells and transformants through a relatively small chamber. The repetition rate and pumping speed can be arranged so that every volume element receives one or, if desired, multiple pulses. Care needs to be taken in the design of the flowthrough chamber to minimize dead volume. Repetitive pulse generators are com-

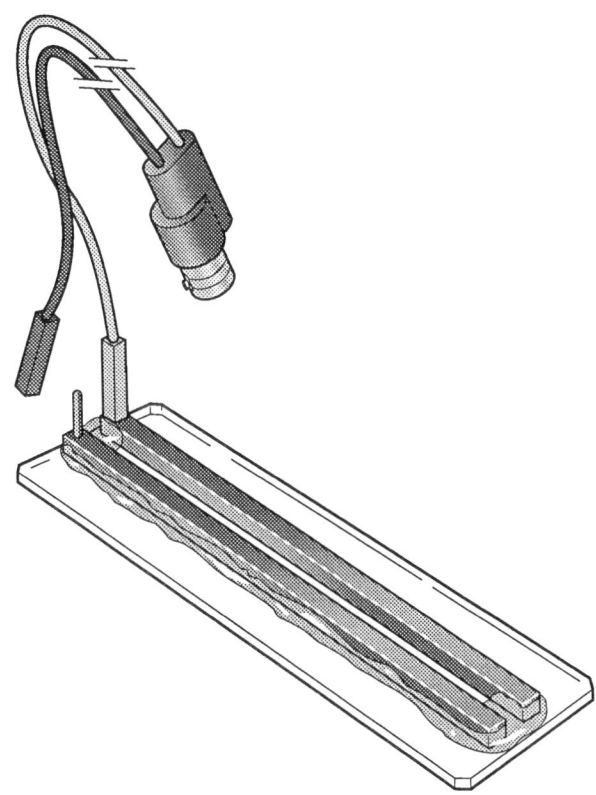

Fig. 7. Rectangular electrodes mounted on a microslide for visual observation, generating homogeneous electric fields.

mercially available for exponential decay, as well as for square-wave form output.

5.4. Chamber Material

Despite an oxide layer present on the surface, aluminum (Al) electrodes appear to give satisfactory results in disposable chambers. The commercially available presterilized cuvets use embedded Al electrodes. Stainless steel (SS) is used more often for reusable chambers. SS can be mechanically cleaned more easily. Gold plating is an option for SS as well as Al, but it appears that the increase in yield does not justify the additional costs. Comparative electrofusion experiments of embryos in either SS or gold-plated chambers did not show a substantial difference in fusion yield. Over 100 fusion experiments were performed using gold-plated electrodes, with a

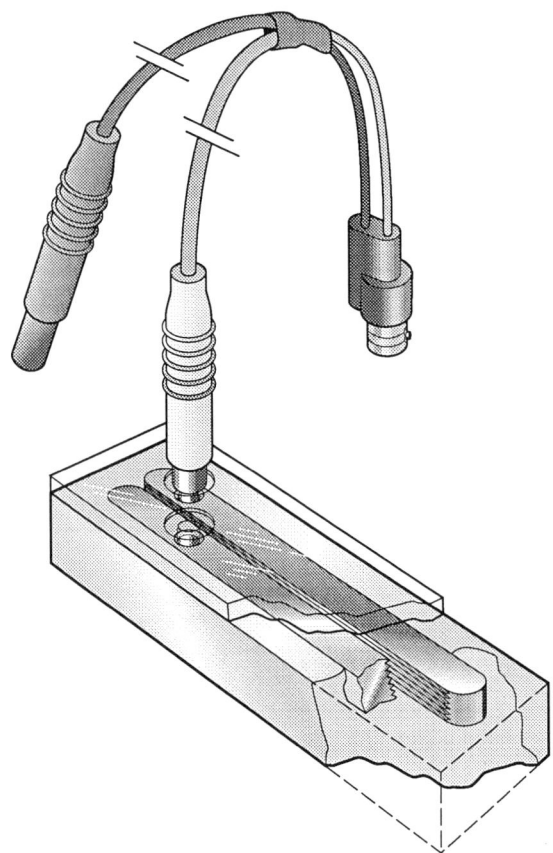

Fig. 8. Intermediate-volume chamber with parallel bar electrodes. The electrodes can be flat to create homogeneous fields or have grooves to create inhomogeneous fields.

fusion yield of >90%; with care, similar results can be achieved with SS electrodes (James M. Robl, U. of Massachusetts, personal communication). A comparison of plant protoplast fusion yields using a large-volume parallel plate chamber made of SS or a gold-plated concentric ring chamber (both for Petri dishes) showed a consistently higher yield for the gold-plated chamber *(10)*. Table 6 shows a survey of commercially available chambers.

6. Measuring ECM Parameters

By following an established protocol, it is generally not necessary to measure the ECM parameters, especially if the same types of generator

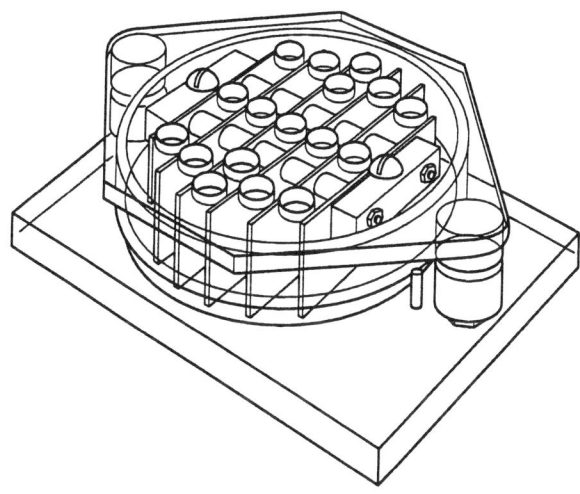

Fig. 9. Petri dish electrodes for large-volume electroporation of mammalian cells in suspension or adherence, and electroporation of fish eggs.

and chamber are used (different generators might give different output voltages for the same charging voltage setting). However, when pursuing new applications with an instrument that does not have built-in monitoring, it is desirable to measure the voltage and pulse actually delivered to the chamber to allow accurate reporting and reproducible performance. A commercial instrument is available to monitor ECM parameters specifically, display them, and print them out (Table 1). A measuring system can be assembled consisting of a digital oscilloscope (bandwidth should be 100 MHz), and a high-voltage probe attenuating the voltage signal 1:1000, with a voltage range up to 3 kV. Commercial generators and chambers are typically constructed so that at no place in the circuit is the high-voltage potential easily accessible. Therefore, adapters need to be placed in line between the generator and the chamber so that a voltage probe can be connected. Before any measurements are performed, the grounding situation must be understood and verified with the manufacturer. Sometimes neither of the two outputs of the generator is at the ground potential of the oscilloscope (which is normally tied to the power line ground), depending on the design of the discharge circuit. It is still possible to perform measurements in this case by disconnecting the oscilloscope from the power line earth/ground, either by inserting an iso-

Table 6
Comparison of Commercially Available Chambers

Chamber types	Manufacturers				
	IBI	Invitrogen	Bio-Rad	BRL	BTX
Cuvets: disposable	3	3	3	3	3
Cuvets: reusable, homogeneous field	3				2
Cuvets: reusable, divergent field	1				1
Microslides homogeneous field					2
Microslides divergent field					2
Meander					1
Flat electrode divergent field					1
Petri dish electrodes					2
96-Well plate electrode					1
Flowthrough					2
Sandwich					2
In vivo electrodes					3

[a]The numbers indicate available variations, typically in the gap size.

lation transformer between the line and the oscilloscope or by disconnecting the oscilloscope ground lead to the power line with an insulating plug. These plugs can be recognized by regular three prongs on one side and only two receptacles on the other side with the earth/ground wire separate, which should not be connected. During the pulse, the chassis of the oscilloscope will attain a potential difference to the laboratory ground and should not be touched. These kinds of measurements obviously are hazardous, and should be performed with extreme care and only by trained personnel. If it is desirable to measure both the voltage output and the current, a convenient, contactless way is to route one lead to the chamber through a current transformer. Several manufacturers provide these elements (e.g., Pearson Electronics Inc. [Palo Alto, CA], Current Transformer Model Nr. 411). The most important specifications to verify the usefulness of a current transformer are the peak current capability (e.g., 5000 A for the 411) and the limit of the product of current × pulse length (e.g., 0.2 A · s for the 411) to avoid saturation of the current transformer before the pulse has passed completely. Measuring current and voltage allows one to determine the chamber resistance as a function of time from Ohm's law, $R = V/I$. Through the geometry of the chamber, the specific resistivity of the cell/medium suspension as a function of time can then be determined, which might be of interest for biophysical investigations of the ECM process, because lysis of cells results in an increase of the medium conductivity.

Acknowledgments

I want to thank my colleagues for helpful comments, and especially Linda Hull for editing the manuscript.

References

1. Tsong, T. Y. and Tomita, M. (1993) Selective B lymphocyte-myeloma cell fusion. *Methods in Enzymol.* **220,** 238–246.
2. Pohl, H. A. (1978) *Dielectrophoresis.* Cambridge University Press, London.
3. Meilhoc, E., Masson, J.-M., and Teissie, J. (1990) High efficiency transformation of intact yeast cells by electric field pulses. *Biotechnology* **8(3),** 223–227.
4. Takahashi, M., Furukawa, T., Saito, H., Aoki, A., Koike, T., Moriyama, Y., Shibata, A. (1991) Gene transfer into human leukemia cell lines by electroporation: experiments with exponentially decaying and square wave pulse. *Leukemia Res.* **15(6),** 507–513.
5. Saunders, J., Rhodes, S. C., and Kaper, J. (1989) Effects of electroporation profiles on the incorporation of viral RNA into tobacco protoplasts. *Biotechniques* **7(10),** 1124–1131.
6. Xie, T. and Tsong, T. (1992) Study of mechanisms of electric field-induced DNA transfection III, Electric parameters and other conditions for effective transfection. *Biophys. J.* **63,** 28–34.
7. Hofmann, G. H. (1989) Cells in electric fields—physical and practical electronic aspects of electro cell fusion and electroporation, in *Electroporation and Electrofusion in Cell Biology* (Neumann, E., Sowers, A., and Jordan, C., eds.), Plenum, New York, pp. 389–407.
8. Nagata, K. and Imai, H. (1992) The difference of electro fusion rate for the pronuclear transplantation of mouse eggs between three different electro generators. The 7th Eastern Japan Animal Nuclear Transplantation Research Conference.
9. Chang, D. C. (1989) Cell fusion and cell poration by pulsed radio-frequency electric fields, in *Electroporation and Electrofusion in Cell Biology* (Neumann, E., Sowers, A., and Jordan, C., eds.), Plenum, New York, pp. 215–227.
10. Tekle, E., Astumian, R. D., and Chock, P. B. (1991) Electroporation by using bipolar oscillating electric field: an improved method for DNA transfection of NIH 3T3 cells. *PNAS* **88,** 4230–4234.
11. Motumura, T., Akihama, T., Hidaka, T., and Omura, M. (1993) Conditions of protoplast isolation and electrical fusion among citrus and its wild relatives, in *Techniques on Gene Diagnosis and Breeding in Fruit Trees* (Hayashi, T., et al., eds.), FTRS, Japan, pp. 153–164.

PART II

ELECTROPORATION PROTOCOLS

CHAPTER 4

Electroporation
of *Agrobacterium tumefaciens*

Amke den Dulk-Ras and Paul J. J. Hooykaas

1. Introduction

Agrobacterium tumefaciens is a soil bacterium that causes tumors on dicotyledonous plants. Virulent strains harbor a large plasmid, the Ti (tumor-inducing) plasmid, which is involved in tumorigenesis. A small segment of this plasmid, the T-DNA, is transferred to the plant cell and becomes integrated into one of the chromosomes in the nucleus. The T-DNA contains genes for the production of phytohormones viz an auxin and a cytokinin. Therefore, expression of the T-DNA in the plant cell leads to tumor formation (for review, *see* ref. *1*). Deletion of the *onc* genes within the T region of the Ti plasmid results in nononcogenic strains. However, if the 24-bp border repeat, which surrounds the T region in the Ti plasmid, is kept intact, the mutated T-DNA is still delivered to plant cells by *Agrobacterium*. This natural plant vector system is used for the genetic engineering of plants (for review, *see* ref. *2*). If genes are added to the T region of the Ti plasmid, these are cotransferred to the plant cell. An important finding was that separating the T region from the remaining part of the Ti plasmid did not prevent transfer of the T-DNA to the plant cell *(3)*. On the basis of this principle, the binary vector system was developed. Binary vectors are wide host-range plasmids that are maintained by both *E. coli* and *A. tumefaciens* and contain an artificial T region into which genes of interest can be cloned. Traditionally,

From: *Methods in Molecular Biology, Vol. 55: Plant Cell Electroporation and Electrofusion Protocols* Edited by: J. A. Nickoloff Humana Press Inc., Totowa, NJ

cloning with binary vectors is done in *E. coli.* The resulting vector is then introduced into an *A. tumefaciens* helper strain for delivery of the T-DNA to plant cells.

The efficient introduction of plasmids into *A. tumefaciens* is thus of great practical importance for plant molecular biology. There are several methods for introducing plasmid DNA into *A. tumefaciens.* The classical method is conjugative transfer by triparental mating *(4).* Although good results are obtained with this method, the procedure is very time-consuming. Alternatively, plasmids may be introduced into *A. tumefaciens* via transformation. However, calcium chloride treatment, generally used for transformation of *E. coli,* is not effective for *A. tumefaciens.* A freeze–thaw treatment *(5)* is effective, but the transformation efficiency is low (maximally 10^3 transformants/μg DNA).

Electroporation can be used to introduce plasmids efficiently into *A. tumefaciens.* Transformation frequencies of up to 10^7 transformants/μg DNA have been reported *(6–10).* After electroporation, transformants are selected by plating the pulsed cell–DNA mixture on selective plates. Owing to the possible growth of spontaneous antibiotic resistant mutants, it is necessary to check for the presence of the electroporated plasmid in the *A. tumefaciens* cell by performing a small-scale plasmid isolation *(6,11).* In this chapter, detailed protocols are described for electroporation as well as plasmid isolation. In Section 4., various applications of electroporation for gene transfer to *A. tumefaciens* are described (*see* Notes 5 and 6).

2. Materials
2.1. Bacterial Strains and Plasmids

A commonly used bacterial strain (LBA288) is a rifampicin resistant derivative of *A. tumefaciens* strain C58 that is cured of its Ti plasmid *(12).* Other derivatives of LBA288 include strain LBA1100, containing the spectinomycin resistant (Sp^r) Ti helper plasmid pAL1100 *(13)* and strain LBA1143, containing the carbenicillin resistant (Sp^{r}Cb^r) Ti helper pAL1143 *(13).* Strain LBA4404, containing the streptomycin resistant (Sm^r) Ti helper plasmid pAL4404 *(3),* is a derivative of wild-type Ach5 (*see* Note 4). Kanamycin resistance (Km^r) can be conferred by DNA of the 7-kbp binary vector pSDM14 *(14)* and the 11-kbp vector pBin19 *(15).*

2.2. Medium and Antibiotics Used
for Selection of Transformed Cells

1. LC medium: 10.0 g tryptone (Difco, Detroit, MI), 5.0 g yeast extract, 8.0 g NaCl, distilled water to 1 L. Sterilize by autoclaving for 20 min at 120°C. When indicated, supplement medium with 0.1% glucose. For solid medium in plates, add 1.8% agar before sterilization. Agar medium can be stored in liquid form for up to 1 wk in a 55°C incubator. When stored for a longer period of time, the medium must be kept in a solid form. It can then be liquified with the aid of a microwave oven before use.
2. Rifampicin: 10 mg/mL dissolved in methanol. This solution can be stored at 4°C for at least 6 mo.
3. Spectinomycin: 50 mg/mL in sterile H_2O.
4. Carbenicillin: 50 mg/mL in sterile H_2O.
5. Kanamycin: 50 mg/mL in sterile H_2O—for use in bacterial selection plates: No further sterilization of the drugs is needed. Solutions remain stable for at least 6 mo at −20°C.
6. Selection plates: Liquid LC agar medium, supplemented with the required antibiotics to final concentrations of 20 µg/mL rifampicin, 100 µg/mL kanamycin, 100 µg/mL spectinomycin, or 75 µg/mL carbenicillin. For LBA4404, however, carbenicillin is used at 10 µg/mL.

2.3. Solutions for Electroporation

1. Washing solution: 1 mM HEPES, adjust to pH 7.0 with NaOH. Autoclave 20 min at 120°C.
2. Electroporation solution: 10% glycerol in distilled H_2O. Filter-sterilize through a 0.22-µm membrane.
3. SOC medium: 2% bacto-tryptone, 0.5% yeast extract, 10 mM NaCl, 2.5 mM KCl, 10 mM MgSO$_4$, 10 mM MgCl$_2$, 20 mM glucose.

2.4. Solutions
for Small-Scale Plasmid Isolations

1. Solution 1: 50 mM glucose, 10 mM EDTA, 25 mM Tris-HCl, pH 8.0, 4 mg/mL lysozyme. Add lysozyme just before use.
2. Solution 2: 1% SDS, 0.2N NaOH. This solution must be freshly prepared from stock solutions of 20% SDS and 4N NaOH.
3. Solution 3: 1 mL H_2O-saturated phenol plus 15 µL 4N NaOH.
4. Solution 4: 3M sodium acetate, adjust to pH 4.8 with acetic acid.
5. Phenol-chloroform: 1 vol of 100 mM Tris-HCl, pH 8.0, saturated phenol plus 1 vol of chloroform:isoamylalcohol (24:1).
6. TE buffer: 10 mM Tris-HCl, 1 mM EDTA, pH 8.0.

3. Methods

3.1. Preparation
of A. tumefaciens Cells for Electroporation

1. Inoculate an LC plate with bacteria, and incubate for 3 d at 29°C.
2. Use a loop to transfer bacteria into 2 mL of LC medium, and incubate at 29°C for 6 h with agitation.
3. Inoculate 100 mL of LC medium, supplemented with 0.1% glucose, with 100 µL of the preculture. Grow the cells overnight at 29°C with vigorous shaking to an OD_{660} of 1.0–1.5 (*see* Note 1).
4. Chill the culture on ice for 15 min, and harvest the cells by centrifugation in a cold rotor at 4000g for 20 min.
5. Resuspend the pellet in 10 mL of 1 mM HEPES, pH 7, and centrifuge as above. Repeat this washing step three times.
6. Wash the pellet in 10 mL of 10% glycerol.
7. Resuspend the pellet in a final volume of 500–750 µL of 10% glycerol. The cell concentration should be $1–5 \times 10^{11}$ cells/mL.
8. Distribute the bacterial suspension in 40-µL aliquots, freeze in liquid nitrogen, and store at –70°C. The cells are usable for electroporation for at least a year under these conditions without significant loss of transformation efficiency.

3.2. Electroporation of A. tumefaciens

1. Gently thaw the cells on ice. This takes 10–15 min.
2. Chill the electroporation cuvets with 0.2-cm electrode gap on ice (*see* Note 7).
3. Add 1–5 µL of plasmid DNA (10 ng) to 40 µL of cell suspension, and mix well or use the cells for direct transfer (*see* Note 5).
4. Transfer the mixture to a prechilled electroporation cuvet. Take care that the suspension is in contact with both electrodes of the cuvet.
5. Apply an electric pulse at 2.5 kV, 25 µF, and 200 Ω (*see* Note 2). This should result in a pulse of 12.5 kV/cm with a time constant of approx 4.7 ms.
6. Immediately add 1 mL of SOC medium, and gently but quickly resuspend the cells with a sterile Pasteur pipet.
7. Transfer the cell suspension to a 1.5-mL tube, and incubate at 29°C for 1–1.5 h.
8. Plate 100-µL aliquots of appropriate dilutions on selective medium. Incubate for 3 d at 29°C.

3.3. Small-Scale Plasmid Isolations

1. Inoculate fresh bacteria in 2 mL of LC medium, and incubate overnight at 29°C with agitation.

2. Centrifuge 0.5 mL of the bacterial culture for 5 min at 12,000g, and discard the supernatant.
3. Suspend the pellet by vortexing in 100 µL of solution 1. Incubate 10 min at room temperature.
4. Add 200 µL of solution 2, mix by inverting the tube four times, and incubate for 10 min at room temperature.
5. Add 30 µL of solution 3, and mix by brief and gentle vortexing.
6. Immediately add 150 µL of solution 4, mix by inversion, and incubate at −20°C for 15 min.
7. Centrifuge for 5 min at 12,000g, and transfer the supernatant to a fresh 1.5-mL tube.
8. Add 400 µL of phenol-chloroform, and mix by brief vortexing.
9. Centrifuge for 3 min at 12,000g, and transfer the aqueous (top) layer to a 1.5-mL tube.
10. Add 800 µL of ice-cold 96% ethanol, and mix by inversion.
11. Precipitate the DNA by incubation at −70°C for 15 min.
12. Centrifuge for 10 min at 12,000g, and discard the supernatant.
13. Wash the pellet with 250 µL of 70% ethanol, and dry briefly in a vacuum centrifuge.
14. Dissolve the pellet in 25 µL of dH_2O or TE buffer (*see* Note 3).
15. Use 5 µL of the DNA solution to check plasmid presence in a 0.6% agarose gel, 10 µL for restriction analysis, and 1–10 µL for transformation of *E. coli* or *Agrobacterium*, e.g., by electroporation (*see* Note 4).

4. Notes

1. Some workers prepare electrocompetent cells from overnight cultures with an OD_{660} of 1–1.5 (4), whereas others use early log-phase cultures with an OD_{660} of 0.5 (7). It is possible that cellular competence for transformation is influenced by the growth phase of the bacterial culture. Therefore, we compared electroporation of two different cell suspensions of LBA288. One suspension was derived from cells that had been cultured overnight as described in Section 3. To obtain a suspension of log-phase cells, an overnight culture was diluted 1:100 in LC medium supplemented with 0.1% glucose and incubated further for about 6 h until the cells had reached an OD_{660} of 0.3. Electrocompetent cell suspensions of 1.1×10^{11} and 2.5×10^{10} cells/mL were thus obtained. Forty-microliter aliquots were used for electroporation with 100 ng pSDM14 DNA. Results are shown in Table 1. The total number of transformants was equal for both cell suspensions. However, the number of transformants per surviving cells was almost 10-fold higher for the early log-phase cells. A higher initial cell density of

Table 1
Effect of Growth Phase on Transformation Frequency[a]

Growth phase, OD_{660}	Cell density, cells/mL	Total number of transformants	Number of transformants per surviving cell
1.1	1.1×10^{11}	2.1×10^4	7.2×10^{-6}
0.3	2.5×10^{10}	2.2×10^4	8.1×10^{-5}

[a]LBA288 cells were grown to an OD_{660} of 1.1 (overnight culture) or to an OD_{660} of 0.3 (early log phase). Transformation was carried out by using 100 ng of pSDM14 DNA, a field strength of 12.5 kV/cm, a capacitance of 25 μF, and a resistance of 200 Ω.

early log-phase cells will probably lead to an increase in the total number of transformants that can be obtained. However, it is more convenient to use overnight cultures. For most purposes, cells prepared from overnight cultures are sufficiently competent.

2. The electroporation process is influenced by several parameters, such as field strength and pulse length. For *A. tumefaciens,* the highest transformation efficiencies were obtained at a field strength of 12.5 kV/cm, a capacitance of 25 μF, and resistances of 200–400 Ω, giving pulse lengths of 4.5–9 ms *(6–10)*. To test the influence of the pulse length on the transformation efficiency and the survival of LBA288 cells, this strain was electroporated with pSDM14 DNA using resistances of 100, 200, 400, and 600 Ω. As shown in Fig. 1, the optimal pulse length was reached at 200 Ω (time constant = 4.8 ms).

3. DNA solutions used for electroporation must have a low ionic strength. As ionic strength increases, resistance of the suspension decreases. Excess ionic strength will cause arcing in the cuvet. DNA can be prepared either by CsCl density gradient centrifugation or by Qiagen (Diagen GmbH, Düsseldorf, Germany) isolation methods. DNA isolated by minialkaline lysis procedures is also usable, but the efficiency of electroporation is about 2.5-fold lower than with the other preparations *(10)*. Small cloning vectors (10–12 kbp) and large 200–250 kbp Ti plasmids can be reproducibly transformed into *A. tumefaciens* strains *(6)*. Transformation frequency was found to be linearly related to DNA concentration from 0.1 pg up to 1 μg/standard assay *(9)*. However, a reproducible twofold decrease in frequency was found in our experiments when comparing the use of 10 and 100 ng DNA, respectively.

4. Several different *A. tumefaciens* and *A. rhizogenes* strains have been tested in electroporation experiments. Differences in the transformation efficiencies obtained with different stains were observed *(9)*. It is important to take this into account when choosing an *Agrobacterium* strain for an

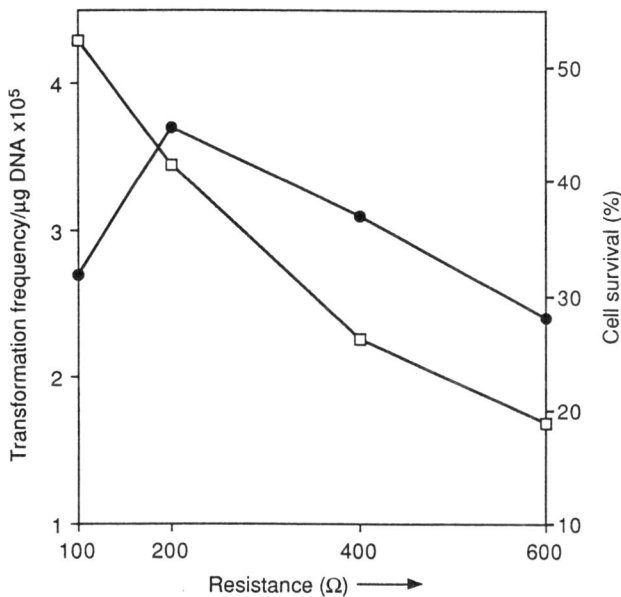

Fig. 1. Influence of pulse time on transformation efficiency and survival. Electroporation was carried out with strain LBA288 using 10 ng of pSDM14 DNA. Pulses were delivered at 12.5 kV/cm, 25 µF, and resistances of 100, 200, 400, and 600 Ω, giving corresponding time constants of 2.4, 4.7, 9.0, and 13.2 ms. Transformation efficiency (●) and survival of cells (□).

Table 2
Comparison of Transformation Frequencies
of *A. tumefaciens* Strains LBA288 (C58) and LBA4404 (Ach5)[a]

Strain	Cell density, cells/mL	Total number of transformants	Frequency in transformants/µg DNA
LBA288	2.5×10^{10}	4.0×10^3	4×10^5
LBA4404	1.8×10^{10}	0.9×10^3	9×10^4

[a]Strains were electroporated under the same conditions: Cells of early log-phase growth stage (OD_{660} of 0.3) were used. Transformation was carried out by using 10 ng pSDM14 DNA, a field strength of 12.5 kV/cm, a capacitance of 25 µF, and a resistance of 200 Ω (time constant 4.8 ms).

investigation. In Table 2, the transformation frequencies are compared for the strains LBA288 and LBA4404, using the protocol given in Section 3.
5. Direct transfer of bacterial plasmid DNA between two *E. coli* strains by electroporation, so-called electrotransfer, was reported by Summers and

Withers *(16)*. Based on these observations, we applied this concept to achieve direct transfer of plasmid DNA between two *A. tumefaciens* strains and from *E. coli* to *A. tumefaciens.* We directly transferred the plasmid pBin19 (Km^r) from LBA288 to LBA1143, which contains an Sp^r, Cb^r helper plasmid, as follows: Both strains were prepared for electroporation as described in the protocol. Twenty-microliter aliquots of both strains were mixed and transferred to a prechilled 0.2-cm electroporation cuvet. The mixture was subjected to a pulse of 12.5 kV/cm, 25 µF, and 200 Ω (time constant 4.8 ms). Immediately after the pulse, the cuvet was incubated on ice for 30 s. Then a second pulse (same settings, time constant 4.6 ms) was applied, followed by addition of 1 mL of SOC medium. After a 1-h incubation at 29°C, 100-µL aliquots of the pulsed cell mixture were spread onto LC-agar plates containing kanamycin (100 µg/mL) and spectinomycin (100 µg/mL). A total of 150 colonies were found to be resistant to both antibiotics. As expected, these colonies also turned out to be resistant to carbenicillin, the other marker of the recipient. In a control experiment where no pulses were applied to the mixture, no transfer was observed. Plasmid presence was examined by electrophoresis of minialkaline DNA isolations. On 0.6% agarose gels, two bands were found representing pBin19 and helper plasmid pAL1143, respectively. This shows that electrotransfer between two *A. tumefaciens* strains had indeed occurred. For the electrotransfer of plasmids from *E. coli* to *A. tumefaciens,* cells of the "empty" strain LBA288 were prepared for electroporation. A fresh colony from a plate of the *E. coli* strain CEL247, which contains the Km^r binary vector pBin19, was gently mixed with the cold suspension of thawed *A. tumefaciens* cells. As described above, two electric pulses were applied. Transformed *A. tumefaciens* strains were selected on plates containing rifampicin (20 µg/mL) and kanamycin (100 µg/mL). In the 24 colonies obtained, the presence of plasmid pBin19 was demonstrated by electrophoresis of DNA preparations. Electrotransfer is a very useful application of electroporation because the DNA isolation step can be omitted.

6. Often it is desirable to insert a gene of interest into the Ti plasmid or the *Agrobacterium* chromosome. This can be accomplished by homologous recombination between an introduced plasmid and the *Agrobacterium* genome. The frequency of such transfer is lower than when using a replication-proficient plasmid. Nevertheless, electroporation can be used for this type of transfer. Here we describe the introduction of a mutated *virG* gene into a helper Ti plasmid. The mutated *virG* gene was cloned in the *E. coli* vector pTZ18R (Cb^r), which cannot replicate in *A. tumefaciens.* DNA (250 ng) of this clone was electroporated into strain LBA1100 *(13),* which contains a helper Ti plasmid with a Sp^r gene. Selection for the Cb

marker after electroporation resulted in growth of several colonies. Southern analysis showed that the clone had integrated into the helper Ti plasmid via homologous recombination by a single crossover event.

7. Although electroporation cuvets are disposable, it is possible to reuse them. The cuvets can be decontaminated by rinsing them twice with 5% dettol, then with distilled H_2O, and finally with 70% ethanol for sterilization. In old cuvets and cuvets that are not cleaned properly, arcing can occur. Replace these cuvets.

References

1. Nester, E. W., Gordon, M. P., Amasino, R. M., and Yanofsky, M. F. (1984) Crown gall: a molecular and physiological analysis. *Ann. Rev. Plant Physiol.* **35,** 387–413.
2. Hooykaas, P. J. J. and Schilperoort, R. A. (1992) Agrobacterium and plant genetic engineering. *Plant Mol. Biol.* **19,** 15–38.
3. Hoekema, A., Hirsch, P. R., Hooykaas, P. J. J., and Schilperoort, R. A. (1983) A binary plant vector strategy based on separation of vir- and T-region of the *A. tumefaciens* Ti-plasmid. *Nature* **303,** 179,180.
4. Ditta, G., Stanfield, S., Corbin, D., and Helinski, D. R. (1980) Broad host range DNA cloning system for gram-negative bacteria: construction of a gene bank of *Rhizobium meliloti. Proc. Natl. Acad. Sci. USA* **77,** 7347–7351.
5. Holsters, M., De Waele, D., Depicker, A., Messens, E., Van Montagu, M., and Schell, J. (1978) Transfection and transformation of *Agrobacterium tumefaciens. Mol. Gen. Genet.* **163,** 181–187.
6. Mozo, T. and Hooykaas, P. J. J. (1991) Electroporation of megaplasmids into Agrobacterium. *Plant Mol. Biol.* **16,** 917,918.
7. Mattanovich, D., Rüker, F., da Camara Machado, A., Laimer, M., Regner, F., Steinkellner, H., Himmler, G., and Katinger, H. (1989) Efficient transformation of *Agrobacterium* spp. by electroporation. *Nucleic Acids Res.* **17,** 6747.
8. Wen-jun, S. and Forde, B. G. (1989) Efficient transformation of *Agrobacterium* spp. by high voltage electroporation. *Nucleic Acids Res.* **17,** 8385.
9. Nagel, R., Elliott, A., Masel, A., Birch, R. G., and Manners, J. M. (1990) Electroporation of binary Ti plasmid vector into *A. tumefaciens* and *A. rhizogenes. FEMS Microbiol. Lett.* **67,** 325–328.
10. Cangelosi, G. A., Best, E. A., Martinetti, G., and Nester, E. W. (1991) Genetic analysis of *Agrobacterium. Methods Enzymol.* **204,** 384–397.
11. Birnboim, H. C. and Doly, J. (1979) A rapid alkaline extraction procedure for screening recombinant plasmid DNA. *Nucleic Acids Res.* **7,** 1513–1523.
12. Koekman, B. P., Hooykaas, P. J. J., and Schilperoort, R. A. (1980) Localization of the replication control region on the physical map of the octopine Ti plasmid. *Plasmid* **4,** 184–195.
13. Beijersbergen, A., den Dulk-Ras, A., Schilperoort, R., and Hooykaas, P. J. J. (1992) Conjugative transfer by the virulence system of *Agrobacterium tumefaciens. Science* **256,** 1324–1327.

14. Offringa, R. (1992) Gene targeting in plants using the *Agrobacterium* vector system. PhD Thesis, Leiden University, The Netherlands.
15. Bevan, M. (1984) Binary *Agrobacterium* vectors for plant transformation. *Nucleic Acids Res.* **12,** 8711–8721.
16. Summers, D. K. and Withers, H. L. (1990) Electrotransfer: direct transfer of bacterial plasmid DNA by electroporation. *Nucleic Acids Res.* **18,** 2192.

CHAPTER 5

Electroporation of DNA into the Unicellular Green Alga *Chlamydomonas reinhardtii*

Laura R. Keller

1. Introduction

Molecular characterization of complex cellular processes depends on the use of model systems in which a classical genetic approach can be combined with molecular techniques of gene cloning and DNA transfer. Several features of the green alga *Chlamydomonas reinhardtii* make it a suitable model system for studying such cellular processes as motility, photosynthesis, respiration, circadian rhythms, intermediary metabolism, and signal transduction. These features include a haploid genome of small size, simple vegetative and sexual cycles, and the ease of maintaining laboratory cultures in liquid or as single colonies on agar plates. Classical genetic analyses of *Chlamydomonas,* including mutagenesis, linkage, and complementation, have produced a large and detailed genetic map *(1).* Recently developed techniques for efficient transformation *(2–8)* and insertional mutagenesis *(9)* of *Chlamydomonas* nuclear and organelle genomes promise to accelerate the genetic analysis of these and other complex cellular processes.

Several methods have been developed for introducing DNA into the *Chlamydomonas* nuclear genome. The first developed "biolistic" method, which involves bombardment of *Chlamydomonas* cells on agar plates with DNA-coated tungsten particles *(7),* has been used successfully for organelle as well as nuclear transformation. Transformation efficiency is

From: *Methods in Molecular Biology, Vol. 55: Plant Cell Electroporation and Electrofusion Protocols* Edited by: J. A. Nickoloff Humana Press Inc., Totowa, NJ

comparatively low using the biolistic method, and transformants may harbor multiple copies of the introduced DNA, sometimes linked in long tandem arrays *(5,7)*. Using a simpler "glass bead" method, in which cells are vortexed with DNA and glass beads, transformants harboring one or a few copies of the introduced DNA are produced at higher transformation efficiencies *(8)*. However, successful transformation with the glass bead method is complicated by removal of the cell wall, using either *Chlamydomonas* autolysin *(1)* or a cell-wall-deficient genetic mutant strain *(10)*. Although even higher transformation efficiencies of cells with intact walls are obtained using a modification of the glass bead technique in which silicon carbide microfibers replace glass beads *(6)*, transformants generated using silicon carbide microfibers are less well characterized at present, and a toxic byproduct formed with this technique requires care in its disposal.

Using electroporation as a means of gene transfer, cells with intact walls can be transiently or stably transformed with one or a few copies of a DNA molecule *(3)*. Special cell strains or treatments for removal of cell walls are not required, and toxic byproducts are not generated. Electroporation has been used to introduce into other cell types a variety of molecules, including DNA, RNA, oligonucleotides, nucleotides, proteins, drugs, and ions, and potentially can be used in a wide range of experimental applications in addition to gene transfer with *Chlamydomonas*.

Using the electroporation conditions delineated below, uptake and transient expression of exogenous DNA by virtually any *C. reinhardtii* strain can be induced. Generation of stably transformed cell lines requires the use of genetically marked strains and growth in selection conditions, and two selection regimes commonly used for identification of stable *Chlamydomonas* transformants are described below. The following time line outlines the steps for gene transfer into *Chlamydomonas* by electroporation.

1. Prior to electroporation:
   a. One to two weeks prior to electroporation, streak cells from slant cultures onto a stock plate.
   b. Three to five days prior to electroporation, seed liquid cultures of cells for electroporation from stock plates.
   c. Check supplies of plasmid DNAs for electroporation, and prepare new stocks if necessary.

Table 1
Components and Preparation of *Chlamydomonas* Culture Medium[a]

Component	mL Stock/ L of media	Final molarity of media	% of Chemical in stock
1. 10X trace metals[b]	1.0	—	[b]
2. Na citrate · 2H$_2$O	5.0	0.0017	10
3. FeCl$_3$ · 6H$_2$O	1.0	0.00037	1
4. CaCl$_2$ · 2H$_2$O	1.0	0.00036	5.3
5. MgSO$_4$	3.0	0.0012	10
6. NH$_4$NO$_3$	3.0	0.0037	10
7. KH$_2$PO$_4$	1.0	0.00074	10
8. K$_2$HPO$_4$	1.0	0.00057	10

[a]Culture medium M I, pH 6.8: Add components 1–8 to distilled water in the order given. Bring to 1 L with distilled water, and autoclave. Store individual components at 4°C, and sterile medium at room temperature.
[b]10X trace metals: H$_3$BO$_3$ (1000 mg/L), ZnSO$_4$ · 7H$_2$O (1000 mg/L), MnSO$_4$ · H$_2$O (303 mg/L), CoCl$_2$ · 6H$_2$O (200 mg/L), Na$_2$MoO$_4$ · 2H$_2$O (200 mg/L), CuSO$_4$ (40 mg/L).

2. On the day of electroporation:
   a. Label tubes and plates for each sample, and set up the experiment in a sterile hood.
   b. Dry DNAs to be used in electroporation in the appropriate culture tubes.
   c. Concentrate the cells to 1×10^7 cells/mL in growth medium.
   d. Electroporate the cells.
3. After electroporation, process the electroporated cells. If stable transformants are desired, then either plate the cells onto agar plates, or transfer to fresh liquid medium for growth in selective conditions. If transient transformants are desired, then gently agitate the cells on a rotary shaker until they are ready for analysis.

2. Materials

1. Flasks of culture medium (*see* Notes 1, 2, and Table 1).
2. Petri plates of culture medium + 1.5% washed agar (*see* Note 3).
3. *Chlamydomonas* cells for electroporation (*see* Notes 4 and 8).
4. Hemacytometer.
5. 2% Glutaraldehyde for cell fixation prior to counting.
6. DNA for electroporation (*see* Note 5).
7. Centrifugal vacuum evaporator.
8. Selection medium and Petri plates of selection medium + 1.5% agar (*see* Note 8 and Table 1). For each electroporation sample, one sterile conical 15-mL polystyrene tube, one 1.5-mL tube, and three Petri plates of selective medium + 1.5% washed agar are needed.

3. Methods
3.1. Cell-Culture Conditions

1. To maintain *Chlamydomonas* cells on stock plates, streak cells from a slant culture onto a plate containing culture medium M I + 1.5% agar, and grow for 1–2 wk until the plates are moderately green with cells.
2. To start cultures of *Chlamydomonas* for electroporation, remove foil cover, inoculate a starter flask of sterile culture medium with a large loop full of cells from a stock plate, and replace the foil leaving the Pasteur pipet end exposed (*see* Note 2).
3. Grow the cells for 3–5 d with gentle aeration (1–5 bubbles/s) using filtered air to a density of 0.1–1 × 10^6 cells/mL (*see* Note 6).

3.2. Electroporation of Cells

1. At least 1 h before electroporation, arrange all materials and supplies other than the cells and DNA in the hood, and turn on the fan and UV light in the hood for decontamination (*see* Note 7).
2. Aliquot plasmid DNA from concentrated stocks into 1.5-mL tubes, and dry in a centrifugal vacuum evaporator (*see* Notes 5 and 8).
3. Measure the volume of *Chlamydomonas* cells available for electroporation, and determine the cell concentration by counting a sample of fixed cells using a hemocytometer. Calculate the volume of culture medium needed to resuspend the cells at a concentration of 1 × 10^7 cells/mL.
4. Harvest the cells by centrifugation in a sterile conical centrifuge tube for 5 min at 900*g* at 4°C. Decant the used culture medium.
5. Gently resuspend the cell pellet in fresh culture medium at 1 × 10^7 cells/mL.
6. Aliquot 0.5 mL of cells into each tube containing dried plasmid DNA, and mix well by gentle pipeting.
7. Transfer the cell–DNA mixture to an electroporation cuvet (*see* Notes 9 and 10), and place the cuvet in the electroporation chamber.
8. Pulse with 100 V/cm, 25 µF, and 200 Ω (*see* Note 11). Transfer the cuvet to ice, and incubate for 5 min in the electroporation cuvet.
9. Transfer the sample to a 15-mL sterile conical centrifuge tube, wash the cuvet with 0.5 mL of selection medium (*see* Note 8), combine the wash with the sample, and incubate on ice for at least an additional 5 min.
10. After all samples have been through steps 8 and 9, add 5 mL of selection medium to each sample in a 15-mL conical tube, balance, centrifuge at 900*g* for 5 min at room temperature, and decant the supernatant.

3.3. Processing of Samples After Electroporation

1. If cells are to be used in transient-expression assays, resuspend the pellet of electroporated cells in 5 mL of culture medium, and grow for

the desired time interval to allow expression of the exogenous gene (*see* Note 12).

2. If the cells are to be screened for stable transformation after integration of the exogenous DNA, resuspend the cell pellet in 0.35 mL of selection medium, and spread on three plates containing selection medium + 1.5% agar.

3. Grow both liquid cultures and plated cells in alternating light/dark cycles and with aeration of liquid cultures.

4. Notes

1. *Chlamydomonas* cells are grown on a simple medium with defined salts. In general, cell strains are maintained on stock plates containing culture medium + agar. For electroporation, cells are grown in flasks of liquid culture medium with aeration using filtered air. In the absence of an added carbon source, the cells grow photosynthetically, and cultures can be synchronized by growth in light intensities of 140 µE/m^2/s on alternating cycles of 14 h light and 10 h dark. In the presence of a carbon source (usually acetate), the cells grow asynchronously, but to a much higher density.

   Cells to be used for electroporation are grown in either M I of Sager and Granick *(11)* or 1/2 R medium. *See* Table 1 for preparation of *Chlamydomonas* culture medium M I stock and working solutions. Culture medium 1/2 R is prepared by adding 5 mL/L of 2.2M sodium acetate and 2 mL/L each of KH$_2$PO$_4$ and K$_2$HPO$_4$ to culture medium M I.

2. To prepare flasks of liquid growth medium, place 75 mL of medium into a 125-mL Erlenmeyer flask, and plug with a foam disposable plug stopper (available from Baxter Scientific Products [Houston, TX], cat. #T1384) pierced with a cotton-plugged 5¼ in. Pasteur pipet. Wrap the entire top with a piece of aluminum foil to retard desiccation, and sterilize by autoclaving for 20 min.

3. The agar should be washed prior to its use in culture plates to remove water-soluble impurities, which are toxic to *Chlamydomonas* cells. Washing agar improves the plating efficiency of both control and transformed cells as much as 1000-fold over plating efficiency on unwashed agar. To prepare 1 L of culture medium + 1.5% washed agar, mark a 2-L Erlenmeyer flask at the 1-L vol. Weigh 15 g of dry agar into the flask along with ~1.5 L of distilled H$_2$O (dH$_2$O), and stir for 1–2 h at room temperature. Allow the agar to settle for ~30 min, and decant off the dH$_2$O. Add ~1.5 L of fresh dH$_2$O, and stir again for 1–2 h. Wash the agar in this manner at least twice more. Finally, add culture medium components, bring the solution to 1-L vol, cover, autoclave, cool slightly, and pour into sterile Petri plates. One liter of growth medium + 1.5% agar is sufficient for preparation of approx sixty 135 × 100 Petri plates.

4. Obtain strains of *Chlamydomonas* cells and DNAs for transformation from the *Chlamydomonas* Culture Center, Botany Department, Duke University, Durham, NC 27708-0338. Additional information about handling this organism can be found in The *Chlamydomonas* Sourcebook *(1)*.

5. The plasmid DNAs used in transformation by electroporation are the same as those used for transformation by other means. Plasmid DNAs are prepared by the alkaline lysis protocol *(12)*, purified by banding in CsCl using standard protocols *(13)*, precipitated with 0.3M sodium acetate and 70% ethanol (final concentrations), dried, and resuspended in sterile distilled H_2O.

6. Ideally, cells should be growing exponentially and concentrated immediately prior to electroporation.

7. Setting up for the electroporation (labeling tubes and plates, gathering supplies) often takes more time than actually performing the electroporation. Depending on the number of samples to be electroporated, allow 1–3 h for the entire process.

8. For producing stably transformed cell lines, two different selectable markers for transformation, NIT1 or ARG7, are commonly used. In one regime, the NIT1 gene (encoding nitrate reductase) is introduced into nit1 mutant cells, allowing growth of transformed cell lines on medium containing nitrate as the sole nitrogen source. *Chlamydomonas* nit1 cells deficient in nitrate reductase enzyme activity are cotransformed with 1–2 μg of the plasmid pMN24, containing the *Chlamydomonas* nitrate reductase gene as a selectable marker, and 1–20 μg of the experimental plasmid *(7,8)*. To prepare selection medium lacking reduced nitrogen, add 4 mL/L of 1M KNO_3 instead of NH_4NO_3 (*see* Table 1).

   In the other regime, the ARG7 gene (encoding argininosuccinate lyase) complements the arginine-requiring arg7 mutant. *Chlamydomonas* arg7 mutant cells, maintained by growth on culture medium supplemented with 50 mg/L of arginine, are cotransformed with 1–2 μg of the plasmid pARG7.8 and 1–20 μg of the experimental plasmid, and grown in selection medium lacking arginine *(4)*.

9. Electroporation devices from Bio-Rad (Richmond, CA) and Gibco-BRL Life Technologies (Gaithersburg, MD) have been used successfully in our lab.

10. Volumes of at least 0.5 mL should be used during electroporation to prevent arcing between the cuvet electrodes. Arcing is less of a potential hazard using cuvets with an electrode gap of 0.4 cm than 0.2 cm.

11. Optimal electroporation conditions differ for cells with and without cell walls. A single pulse is sufficient for transformation of cells without walls, such as the cell-wall-less mutant cw-15 *(3)*, whereas two pulses are optimal for cells with intact cell walls. To deliver double pulses to walled cells,

deliver the first pulse and record the time constant. Wait 10–15 s, and administer another pulse at the same settings. Record the time constant for this pulse.

12. Samples are assayed for transient expression between 12 and 96 h after electroporation. If postelectroporation conditions require growth or maintenance of cells for periods of only 3–4 h, then precautions to keep the cultures sterile are unnecessary.

References

1. Harris, E. H. (1989) *The* Chlamydomonas *Sourcebook: A Comprehensive Guide to Biology and Laboratory Use.* Academic, San Diego, CA.

2. Boynton, J., Gillham, N., Harris, E., Hosler, J., Johnson, A., Jones, A., Randolph-Anderson, B., Robertson, D., Klein, T., Shark, K., and Sanford, J. (1988) Chloroplast transformation in *Chlamydomonas* with high-velocity microprojectiles. *Science* **240**, 1534–1538.

3. Brown, L., Sprecher, S. L., and Keller, L. R. (1991) Introduction of exogenous DNA into *Chlamydomonas reinhardtii* by electroporation. *Mol. Cell. Biol.* **11**, 2328–2332.

4. Debuchy, R., Purton, S., and Rochaix, J.-D. (1989) The argininosuccinate lyase gene of *Chlamydomonas reinhardtii*: an important tool for nuclear transformation and for correlating the genetic and molecular maps of the ARG7 locus. *EMBO J.* **8**, 2803–2809.

5. Deiner, D. R., Curry, A. M., Johnson, K. A., Williams, B. D., Lefebvre, P. L., Kindle, K. L., and Rosenbaum, J. L. (1990) Rescue of a paralyzed-flagella mutant of *Chlamydomonas* by transformation. *Proc. Natl. Acad. Sci. USA* **87**, 5739–5743.

6. Dunahay, T. C. (1993) Transformation of *Chlamydomonas reinhardtii* with silicon carbide whiskers. *BioTechniques* **15**, 452–454.

7. Kindle, K., Schnell, R., Fernandez, E., and Lefebvre, P. (1989) Stable nuclear transformation of *Chlamydomonas* using the *Chlamydomonas* gene for nitrate reductase. *J. Cell Biol.* **109**, 2589–2601.

8. Kindle, K. L. (1990) High frequency nuclear transformation of *Chlamydomonas reinhardtii*. *Proc. Natl. Acad. Sci. USA* **87**, 1228–1232.

9. Tam, L. W. and Lefebvre, P. L. (1993) Cloning of flagellar genes in *Chlamydomonas reinhardtii* by DNA insertional mutagenesis. *Genetics* **135**, 375–384.

10. Davies, D. and Plaskitt, A. (1971) Genetic and structural analysis of cell wall formation in *Chlamydomonas reinhardtii*. *Genet. Res.* **17**, 33–43.

11. Sager, R. and Granick, S. (1953) Nutritional studies with *Chlamydomonas reinhardtii*. *Ann. NY Acad. Sci.* **56**, 831–838.

12. Birnboim, H. C. and Doly, J. (1979) A rapid alkaline extraction procedure for screening recombinant plasmid DNA. *Nucleic Acids Res.* **7**, 1513–1523.

13. Sambrook, J., Fritsch, E. F., and Maniatis, T. (1989) *Molecular Cloning, A Laboratory Manual,* 2nd ed. Cold Spring Harbor Laboratory, Cold Spring Harbor, NY.

CHAPTER 6

Pollen Electrotransformation in Tobacco

James A. Saunders and Benjamin F. Matthews

1. Introduction

Numerous techniques have been developed to transfer genes into plants to create genetically engineered crops that can tolerate environmental stresses, and to improve productivity and quality. The search for easier, more efficient techniques to transfer genes continues because the efficiencies of current techniques are low and recovering fertile transgenic plants is difficult and time consuming with some plant species.

Stable transformation of plant cells has been achieved using a number of different mechanisms for DNA uptake. Transforming pollen with genetically engineered genes and using this pollen to fertilize flowers to produce genetically engineered seed is one promising research area for obtaining transgenic plants faster and easier than some previous procedures. This transformation approach is beginning to receive more attention because it circumvents the need for tissue culture, which requires extensive facilities for maintenance and manipulation of sterile explants. It also avoids the time-consuming task of regenerating whole plants from transformed protoplasts or plant tissues, which can take several months, require intensive labor, and expensive facilities. Often, because of the many months of tissue culture required to regenerate plants, many of the regenerated, transgenic plants are infertile and produce no seed, thus further delaying the program.

Pollen transformation also bypasses the use of *Agrobacterium tumefaciens,* which is commonly used to produce transgenic plants of tobacco and several other plants. *A. tumefaciens* works well with a number of

From: *Methods in Molecular Biology, Vol. 55: Plant Cell Electroporation and Electrofusion Protocols* Edited by: J. A. Nickoloff Humana Press Inc., Totowa, NJ

plants and is a routine procedure in many laboratories, however it has a limited host range and does not efficiently transform all plant types. Pollen transformation does not require *A. tumefaciens,* therefore is not limited to the *A. tumefaciens* host range, being limited only to the broader range of plants reproducing via pollen. The underdeveloped technology of pollen transformation has the potential to produce a large variety of transgenic plants in species previously difficult to genetically engineer.

The concept of the use of pollen to effect genetic modification of subsequent progeny has been cited in the literature for some time and the term pollen transformation was coined in the 1970s *(1)*. Several researchers have suggested that DNA, when added to pollen in either a solution or a paste, is capable of being taken up and expressed in progeny. For example, De Wet et al. *(2)* and Ohta *(3)* both using corn, indicated that pollen treated with exogenously added DNA could fertilize flowers and produce seed that phenotypically expressed characteristics of the foreign DNA. Pandey *(4,5)* described the sterilization of pollen of *Nicotiana* by X-ray irradiation and successful pollination of flowers with these treated samples. This report provided evidence that pollen transformation may be functional using the denucleated pollen as a DNA vector. In addition, Hess *(1)* described a series of reports using petunia and corn that indicate that some DNA uptake may occur in pollen exposed to exogenously added DNA. Unfortunately, none of these reports proposed any mechanism for introducing the DNA into the pollen, nor did they obtain conclusive molecular evidence that gene transfer actually occurred.

These deficiencies were pointed out by Sanford et al. *(6)* and others who were unable to repeat the results of these pioneer studies and the process of pollen transformation had been left in a state of skepticism. Additional complications were added by Matousek and Tupy *(7)* and Roeckel et al. *(8),* who described very active DNA nucleases that were present on the pollen wall and were released in an active state after pollen germination. These studies suggest that DNA would be degraded within a few minutes if present in a mixture with germinating pollen. As a result of these negative reports, research on pollen transformation declined, waiting for a mechanism to be found to incorporate exogenously supplied DNA into the pollen grain rapidly enough so that it is not degraded by nucleases. Results from our laboratory indicate that electroporation is an effective mechanism for transferring DNA rapidly into germinating pollen.

Our laboratory reported methods for incorporating radiolabeled DNA into germinating pollen by electroporation *(9),* including Southern blot analysis to confirm DNA uptake *(10).* Also, we reported the production of transgenic tobacco plants from pollen containing either the gene encoding β-glucuronidase (GUS) or chloramphenicol acetyltransferase (CAT) *(11).* The transgenic nature of these plants was confirmed by the presence of DNA encoding marker genes as detected by Southern hybridization and by PCR amplification and hybridization. Expression of GUS activity measured by fluorometric assay and the visualization of GUS activity by histological staining provided further evidence that these were transgenic plants. In addition, it was demonstrated that in tobacco endogenous nuclease activity can be reduced to acceptable levels by washing the pollen with fresh media after germination and immediately prior to the electroporation treatment *(12).* Here we report the details of the protocols for pollen electrotransformation.

2. Materials

1. Pollen is collected from tobacco (*Nicotiana gossei* L. Domin) plants grown in the greenhouse under natural light supplemented with fluorescent light to achieve a 16-h photoperiod or from field grown plants. Tobacco pollen is collected in the morning and used the same day; however, we have stored the pollen at $-70°C$ for up to 6 wk without serious decreases in pollen viability.
2. Germination medium (GM; 13): 10% (w/v) sucrose, 1.27 mM Ca(NO$_3$)$_2$, 0.16 mM H$_3$BO$_3$, and 1 mM KNO$_3$, pH adjusted to 5.2 with additional borate.
3. Electroporation equipment: square wave generator (e.g., BTX model 200, BTX Inc., San Diego, CA).
4. The GUS reporter gene construct pBI221 is a 5.7-kbp plasmid (ClonTech, Palo Alto, CA). Linearize the plasmid with EcoRI prior to electroporation to facilitate incorporation.
5. Gibberellic acid (GA$_3$): Dissolve in ethanol as a stock solution of 100 μg/mL and store at $-10°C$ until use.
6. Plant seeds in 5-in. pots containing Metro-Mix 500 (Grace Sierra Horticultural Products Co., Milpitas, CA) and soil mixed 1:1; water daily.
7. Trypan blue solution: 1 mg/mL in 0.6M mannitol.
8. Fluorescein diacetate (FDA): Prepare a stock solution of 1 mg/mL in acetone. Dilute the stock solution (1:5) with 0.6M mannitol for use. The diluted working solution is only stable for 1 h before the dye precipitates out of solution.
9. A fluorescent microscope equipped with excitation wavelength filter of 485 nm and an emission filter of 520 nm.

3. Methods

3.1. Pollen Electroporation

1. Germinate pollen at a concentration of 4 mg/300 µL in GM in a 2-mm disposable electroporation chamber (e.g., Model 620, BTX Inc.) for 1 h at 30°C in a rotary shaker (50 rpm).

2. Change GM immediately prior to the electroporation to remove any endogenous nucleases that may be released from the pollen wall during germination *(7,12,14)*. The GM can be removed by gentle aspiration after either gravity sedimentation of the pollen or after the pollen is pelleted by centrifugation at 50*g* in a horizontal rotor for 1 min.

3. Add the linearized DNA of interest at a final concentration of 10–20 µg/mL, immediately before the electroporation pulse.

4. Typically pollen treatments and controls include:
   a. No electroporation in the presence of DNA;
   b. Electroporation with DNA; and
   c. Electroporation without DNA.

5. Electroporate the pollen by either a single or dual pulse from a square or exponential pulse generator (*see* Note 1). It is best to examine a range of field strengths from 0–10 kV/cm in 0.5-kV increments. For tobacco pollen, a square wave pulse duration of 80 µs and a cuvet with a 1-mm electrode gap can be used. The optimal conditions for transformation of the germinating tobacco pollen with DNA employs a single 8.75 kV/cm, 80 µs square wave pulse. Alternatively, an exponential pulse generator (BTX Model 600) can be used to achieve efficient pollen electrotransformation in germinating tobacco pollen. We tested exponential pulses from 0–10 kV/cm in 0.5-kV increments using a cuvet with a 2-mm electrode gap at a resistance of 360 Ω. With this system, successful incorporation of DNA was accomplished between 4 and 6 kV/cm while maintaining high pollen viability (*see* Note 2). The range of successful pulse field strengths for the exponential pulse generator is slightly narrower than that of the square wave generator (*see* Note 3).

6. Following electroporation, gravity sediment the pollen within the electroporation cuvet for 10 min. Very carefully remove most of the electroporation medium without disturbing the pollen using a pipet.

7. Emasculate recipient flowers by removing the anthers and bagging the inflorescence 4 d before the electroporation treatment to prevent pollination. Emasculation of flowers prior to electroporation greatly enhances the production of viable seed produced through fertilization with electroporated pollen (*see* Note 4).

8. Pollinate the stigmas of emasculated flowers using the concentrated pollen slurry. Pollinate flowers by pipeting 10–20 μL of the pollen onto the top of the stigma. Rebag the flower for seed production. In some species, such as corn or alfalfa, it may be easier to scoop the pollen slurry out of the electroporation cuvet with a spatula. In any case, use extreme care to prevent damaging the fragile pollen tubes.

9. Collect seeds from pollen-treated flowers and determine seed number, viability, and the number of plants showing positive expression of the introduced trait. Putative transformants that yield a positive expression assay can be verified by Southern hybridization and/or PCR amplification of the GUS coding region of the pBI221 plasmid *(11)*.

3.2. Seed Processing and Planting

1. Surface sterilize seeds obtained from electroporated pollen-treated flowers with 10% (v/v) commercial bleach. Incubate seeds in 200 μL of GA$_3$ at a concentration of 10 μg/mL for 10 min to release them from dormancy.

2. Plant the seeds in Metro-Mix 500 and soil on a misting bench in the green house and grow until the plants are large enough to be transplanted individually into 5 in. clay pots.

3.3. Optimizing Electroporation Conditions with Cytochemical Stains: Viability and Uptake Assay

It is desirable to examine a range of electroporation conditions when pollen from different species or varieties are used. This can be quite tedious when doing expression assays, particularly with plants produced from seed of treated pollen that had been electroporated weeks or months earlier. The combination of two cytochemical stains, trypan blue and fluorescein diacetate (FDA), can be used to rapidly determine the optimal electroporation conditions for DNA uptake while maintaining pollen viability *(15)*.

1. Electroporate 200 μL aliquots of germinated pollen with 20 μL of trypan blue solution.

2. Immediately observe the cells under a bright field microscope and score for the uptake of the blue dye. Blue color indicates that the cell membrane of the protoplasts has been permeated by the electroporation pulse, however, it does not distinguish between cells that are alive and those that have been killed by the electroporation treatment.

3. An aliquot of cells from the same population should be electroporated without trypan blue and stained for viability with an equal volume of 0.1 mg/mL

FDA in 0.6*M* mannitol after a 1-h incubation at room temperature. Bright yellow-green fluorescence of cells indicates a positive viability assay with a fluorescent microscope as described previously *(16,17)*. Examine at least three replicates of 100 cells by light microscopy and score for both viability using FDA and uptake with trypan blue. The working range for electroporation conditions is that where trypan blue uptake has occurred, but without excessive loss of cell viability. Generally this is the point on the graph where the two lines intersect.

4. Notes

1. In general, two different DC high voltage pulse wave forms can be utilized to transform pollen, the square wave pulse and the exponential wave pulse. Using the square wave pulse, both the amplitude and the duration of the pulse can be accurately controlled. With the exponentially decaying pulse, the amplitude of the wave form can be accurately controlled, however, the duration of the pulse can only be modified in a general manner. As in the case of plant protoplasts *(18,19)*, animal cells *(20–22)*, and yeast *(23)*, the success and efficiency of introducing DNA or RNA into the germinating pollen by electroporation depends on several important variables *(9,10,24)*, including the pulse field strength, the pulse duration, the resealing time of the pores introduced into the cell membrane, and the concentration of pollen and DNA in the electroporation medium *(25)*.

2. The field strength of the pulse is controlled by two components: the applied current and the electrode gap. To have an effective electroporation pulse, minimal threshold levels for both the pulse duration and the pulse field strength must be exceeded. Our data suggest that the field strength of the pulse interacts with pulse duration such that, over a limited range, one variable may be increased as the other is decreased and a reversible pore may still be induced. Liang et al. *(26)* have suggested that pore induction, size, and frequency are controlled by pulse height, whereas the length of time the pores remain open is controlled by pulse duration. Benz and Zimmerman *(27)*, have indicated that different cell types require different field strengths to induce pores because of differences in membrane composition and osmotic properties. Pulses that do not meet minimum field strength thresholds of durations may not induce any pore formation, and excessive field strengths lead to irreversible breakdown of the cell membranes.

3. We recommend testing a broad range of pulse field strengths when attempting the electroporation of a new population of cells. This is to determine exactly the pulse field strength sufficient to cause DNA uptake and to determine if the pulse field strength is causing unacceptable damage to the cell viability. Although we do not recommend the procedure, there is

an understandable tendency on the part of many investigators to electroporate their cell lines at published pulse field strengths without checking uptake and viability. Without an accurate knowledge of the effects of the electroporation pulse on the viability of the cells being used, a square wave pulse of <100 µs may yield more "forgiving" positive results over a broader range of pulse field strengths.

4. The emasculation procedure was necessary to prevent pollination of the flower by nontreated pollen. Typically, emasculation was performed by plucking the anther sacs from the flower before flower anthesis, but definitely before the anthers opened. There appeared to be a stimulatory response to seed set with emasculation several days before pollination. For example, emasculation of the recipient flowers in *Nicotiana gossei* prior to pollination with electroporated pollen resulted in a 10-fold increase in the number of seeds formed in the seed capsule *(9)*.

References

1. Hess, D. (1987) Pollen-based techniques in genetic manipulation. *Inter. Rev. Cytol.* **107,** 367–395.
2. De Wet, J. M. J., Bergquist, R. R., Harlan, J. F., Brink, D. E., Cohen, C. E., Newell, C. A., and De Wet, A.-E. (1985) Exogenous gene transfer in maize *(Zea mays)* using DNA-treated pollen, in *Experimental Manipulation of Ovule Tissues* (Chapman, G. P., Mantell, S. H., and Daniels, R. W., eds.) Longman, London, pp. 197–209.
3. Ohta, Y. (1986) High-efficiency genetic transformation of maize by a mixture of pollen and exogenous DNA. *Proc. Natl. Acad. Sci. USA* **83,** 715–719.
4. Pandey, K. K. (1978) Gametic gene transfer in *Nicotiana* by means of irradiated pollen. *Genetica* **49,** 53–69.
5. Pandey, K. K. (1980) Further evidence for egg transformation in *Nicotiana*. *Heredity* **45,** 15–29.
6. Sanford, J. C., Skubik, K. A., and Reisch, B. I. (1985) Attempted pollen-mediated plant transformation employing genomic donor DNA. *Theor. Appl. Genet.* **69,** 571–574.
7. Matousek, J. and Tupy, J. (1983) The release of nucleases from tobacco pollen. *Plant Sci. Lett.* **30,** 83–89.
8. Roeckel, P., Heizmann, P., Dubois, M., and Dumas, C. (1988) Attempts to transform *Zea mays* via pollen grains, effect of pollen and stigma nuclease activities. *Sex. Plant Reprod.* **1,** 156–163.
9. Abdul-Baki, A. A., Saunders, J. A., Matthews, B. F., and Pittarelli, G. W. (1990) DNA uptake by electroporation of germinating pollen grains. *Plant Sci.* **70,** 181–190.
10. Matthews, B. F., Abdul-Baki, A. A., and Saunders, J. A. (1990) Expression of a foreign gene in electroporated pollen grains of tobacco. *Sex. Plant Reprod.* **3,** 147–151.
11. Smith, C. R., Saunders, J. A., Van Wert, S., Cheng, J., and Matthews, B. F. (1994) Expression of GUS and CAT activities using electrotransformed pollen. *Plant Sci.* **104,** 49–58.

12. Van Wert, S. L. and Saunders J. A. (1992) Reduction of nuclease activity released from germinating pollen under conditions used for pollen electrotransformation. *Plant Sci.* **84,** 11–16.

13. Dickinson, D. B. (1968) Rapid starch synthesis associated with increased respiration in germinating lily pollen. *Plant Phys.* **43,** 1–8.

14. Matousek, J. and Tupy, J. (1984) Purification and properties of extracellular nuclease from tobacco pollen. *Bio. Plantarum* **26,** 62–73.

15. Saunders, J. A., Lin, C. H., Cheng, J., Tsengwa, N., Lin, J. J., Smith, C. R., McIntosh, M., and Wert, S. V. (1994) Rapid optimization of electroporation conditions for plant cells, protoplasts, and pollen. *Mol. Biotechnol.* (in press).

16. Saunders, J. A., Roskos, L. A., Mischke, B. S., Aly, M., and Owens, L. D. (1986) Behavior and viability of tobacco protoplasts in response to electrofusion parameters. *Plant Physiol.* **80,** 117–121.

17. Abdul-Baki, A. A. (1992) Determination of pollen viability in tomatoes. *J. Am. Soc. Hort. Sci.* **117(3),** 473–476.

18. Fromm, M., Taylor, L. P., and Walbot, V. (1985) Expression of genes transferred into monocot and dicot plant cells by electroporation. *Proc. Natl. Acad. Sci. USA* **82,** 5824–5828.

19. Fromm, M. E., Taylor, L. P., and Walbot, V. (1986) Stable transformation of maize after gene transfer by electroporation. *Nature* **319,** 791–793.

20. Neumann, E., Schaefer-Ridder, M., Wang, Y., and Hofschneider, P. H. (1982) Inhibition of gene expression in plant cells by expression of antisense RNA. *EMBO. J.* **1,** 841–845.

21. Wong, T. K. and Neumann, E. (1982) Electric field mediated gene transfer. *Biochem. Biophys. Res. Comm.* **107,** 584–587.

22. Potter, H., Weir, L., and Leder, P. (1984) Enhancer-dependent expression of human κ immunoglobulin genes introduced into mouse pre-B lymphocytes by electroporation. *Proc. Natl. Acad. Sci. USA* **81,** 7161–7165.

23. Weber, H., Forester, W., and Jacob, H. E. (1981) Parasexual hybridization of yeasts by electric field stimulated fusion of protoplasts. *Curr. Genet.* **4,** 165,166.

24. Saunders, J. A., Matthews, B. F., and Van Wert, S. L. (1991) Pollen electrotransformation for gene transfer in plants, in *Guide to Electroporation and Electrofusion* (Chang, D. C., Chassy, B. M., Saunders, J. A., and Sowers, A. E., eds.), Academic, San Diego, CA, pp. 227–247.

25. Saunders, J. A., Matthews, B. F., and Miller, P. D. (1989) Plant gene transfer using electrofusion and electroporation, in *Electroporation and Electrofusion in Cell Biology* (Neumann, E., Sowers, A., and Jordan, C., eds.) Plenum, New York, pp. 343–354.

26. Liang, H., Purucker, W. J., Stenger, D. A., Kubiniec, R. T., and Hui, S. W. (1988) Uptake of fluorescence-labeled dextrans by 10T 1/2 fibroblasts following permeation by rectangular and exponential electric field pulses. *BioTechniques* **6,** 550–558.

27. Benz, R., Zimmermann, U., and Wecker, E. (1981) High electric fields effects on the cell membranes of *Halicystis parvula*: a charge-pulse study. *Planta* **152,** 314–318.

CHAPTER 7

Electroporation of Tobacco Leaf Protoplasts Using Plasmid DNA or Total Genomic DNA

Patrick Gallois, Keith Lindsey, and Renee Malone

1. Introduction

Direct gene transfer into tobacco leaf protoplasts via electroporation is used in many laboratories for transient-expression studies. Protoplasts in general are a convenient model to study events that occur *in planta* rapidly. Transient expression provides answers in a matter of days instead of the several months that would be needed if the same experiment was carried out with transgenic plants. Examples include: elicitor regulation of a plant defense gene *(1)*, UV induction of chalcone synthase *(2)*, induction of an Adh1 promoter by low concentrations of oxygen *(3)*, hormone response *(4,5)*, transactivation of a promoter by cotransfection with two plasmids, one encoding a transcription factor and the other one carrying its target promoter in front of a reporter gene *(6)*, homologous recombination *(7)*, and nuclear targeting *(8,9)*. However it must be noted that although in the above examples transient expression is mimicking what can be observed in stable transformants, there are cases reported where there are noticeable differences between the two approaches. For example, the strictly seed-specific gene encoding soybean β-conglycinin is expressed in tobacco mesophyll protoplasts, but the sequences responsible for that expression are different from those controlling expression in stable-transformation experiments *(10)*. The same situation has been suggested for the promoter of the pathogenesis-related 1a protein gene *(11)*.

From: *Methods in Molecular Biology, Vol. 55: Plant Cell Electroporation and Electrofusion Protocols* Edited by: J. A. Nickoloff Humana Press Inc., Totowa, NJ

Direct gene transfer into tobacco protoplasts for stable transformation is used only for specific applications involving rare events, such as homologous recombination, genomic DNA rearrangement, or direct gene transfer using total genomic DNA. If only a few hundred independent transformants are required, tobacco cells can be easily stably transformed using *Agrobacterium tumefaciens.*

We found that transfection of DNA into tobacco protoplasts by electroporation compared with polyethylene glycol (PEG)-based methods gives equally high efficiency, is very convenient, and requires fewer steps. The regeneration steps of the protocol presented here are derived from Negrutiu et al. *(12)* and Installé et al. *(13)*. In our hands, it has worked well for promoter analysis and stable-transformation experiments *(14,15)*. The protocol can be outlined as follows: The leaf tissue is digested overnight by a mixture of a cellulase/pectinase; the protoplasts are then purified from leaf debris by three purifications on a sucrose gradient and resuspended in electroporation buffer. For the electroporation itself, the electrical parameters given here work well in our hands. However, we also provide an example of optimization of the electrical parameters. Note that the electrical parameters for stable transformation or transient expression are slightly different. After electroporation, the culture conditions also differ between the transient-expression and the stable-transformation protocols. For transient expression, the protoplasts are cultivated in liquid medium for 24–36 h before assaying the reporter gene. For stable transformation, the protoplasts are embedded in low-melting agarose and floated on regeneration medium supplemented with a selective agent.

To isolate unknown genes encoding traits that are monogenic and dominant, several strategies are possible. One is map-based cloning *(16)*, which requires a fairly complete genetic map with DNA markers. Another is screening for a loss of function owing to a gene disruption. The loss of function can be caused by the insertion of a T-DNA or a transposon (if available in the chosen species) *(17,18)*. These approaches necessitate the production of a large population of individuals bearing insertions at different loci and have been used mainly with model species. A third strategy is to screen for a gain of function by the introduction of genomic DNA from a donor genome into a cultivar or a species lacking the trait of interest (gene rescue or shotgun cloning). Model experiments have been carried out where genomic libraries have been transferred into plants via *Agrobacterium* with a view to recover a gene of interest *(19–23).* These studies showed that the approach was technically possible *(21,22),* but in practical terms the large size of genomes other than that of *Arabidopsis* limits their general applicability because of the large effort in producing the required transformants. Simoens et al. *(19)* calculated that 17×10^6 transformants must be generated to transfer a particular DNA sequence from the tobacco genome into a host plant via *Agrobacterium* transformation.

We have demonstrated that it is possible to use direct gene transfer (electroporation or PEG-mediated) of genomic DNA to identify and possibly isolate selectable genes from plant genomes larger than *Arabidopsis* *(24).* From the data we have obtained in two experiments involving different selectable genes, one using *Arabidopsis* as a donor plant and the other tobacco, it can be estimated that for a 100,000-kbp genome, it is possible to obtain 5–8% of the transformants that would express the desired gene. This value is valid for a donor plant homozygous for the gene of interest; if heterozygous, the value should be reduced twofold or more depending on the ploidy level of the species. This frequency can be recalculated for larger genome sizes by dividing the frequency by a multiple of 100,000 kbp corresponding to the genome size of the species used as donor. Genome sizes of many plants have been compiled by Arumuganathan and Earle *(25),* Bennett et al. *(26),* and Bennett and Smith *(27).* Roughly, it can be estimated that about 2000–3000 transformants are sufficient to transfer a dominant gene with a tag linked to it from a given species to a cultivated species. This frequency allows the application of a selection or screening scheme at the whole-plant level to detect the transfer of a dominant trait from a donor species to a crop species, even with a low transformation frequency.

Key factors in the successful use of this strategy with crop plants are:

1. A selection scheme for the trait of interest that should be dominant and monogenic. A trait encoded by two genes could also be transferred, but at a lower frequency than a monogenic trait.
2. A transformation method as efficient as possible. The transformation frequency determines the number of cells to be treated for DNA uptake in order to generate the required number of transformants.
3. The transfer of the donor genomic DNA along with a known selectable marker (e.g., on a plasmid), such as kanamycin, hygromycin, or Basta resistance. This allows the estimation of the number of transformants pro-

duced and therefore the expected number of plants transformed with the gene of interest. The treated protoplasts can be first selected for the known selectable marker and the transformant population obtained can be rescreened for the gene of interest.

4. Presence of a known selectable marker in the genome of the donor plant, preferably a single copy per haploid genome, to assess the success of the transfer of the genomic DNA. The marker should be easily scored, such as a resistance/tolerance marker or a visual marker. It could be either a wild-type gene from the donor plant transferred into a mutant host plant, a dominant mutation, or a chimeric gene introduced by transformation in the donor plant.

Because of the large amount of DNA introduced in each transformant (estimated to be around 20,000 kbp with our experimental conditions), cloning involves using individual primary transformants to generate secondary transformants. For this, the DNA from a single primary transformant is reintroduced at random into protoplasts and transformants obtained and screened. The secondary transformants positive for the desired trait are then used for DNA extraction and cloning. The dominant gene could be rescued from the transformed material by cloning or plasmid rescue, both of which have been successfully used for gene rescue in animal research *(28–30)*. Alternatively, if the introduced gene is not expressed in the untransformed line, a cDNA library could be constructed from the transformed line and the relevant cDNA isolated by subtraction with cDNAs from an untransformed line *(31)*. Because plant cells are totipotent, whole-plant regeneration and genetic linkage analysis are possible. The sequences containing the gene of interest are identified at the molecular level by the tag (i.e., kanamycin resistance) used to generate primary transformants. The identification is confirmed by transformation of the cloned DNA fragment in the host plant. Modifications to the basic technique presented here may involve ligating the donor DNA to the plasmid before transformation or cloning the donor DNA in a λ vector before transfer. The λ vector should in that case harbor a selectable gene. The frequency of expected transformants should be recalculated to take into account the presence of λ arms in the DNA transferred to plants.

The protocol presented here has been used for direct gene transfer of genomic DNA extracted from an *Arabidopsis* mutant into tobacco leaf protoplasts *(24)*. The genomic DNA is cotransfected with a plasmid carrying a kanamycin resistance gene. The transfected protoplasts are first selected for resistance to kanamycin and then screened for chlorsulfuron tolerance, which is here the trait of interest. Using a different host species will likely require adjustment of the electroporation parameters and the regeneration conditions.

2. Materials
2.1. Antibiotic Solutions

1. Cefotaxime: 200 μg/mL in water (Roussel Uclaf; sold by Sigma, St. Louis, MO, or preferably local chemist). Filter-sterilize (0.24-μm filter), and prepare just before use.
2. Kanamycin (Sigma): 100 μg/mL, filter-sterilize (0.24-μm filter), store at −20°C.
3. Chlorsulfuron: 200 μg/mL in 10 m*M* potassium phosphate buffer, pH 7.5. Filter-sterilize (0.24-μm filter), and store frozen.

2.2. Protoplast Culture Media

1. Tobacco shoot culture medium: 1X Murashige and Skoog salts and vitamins *(32)* (Flow Laboratories Inc., McLean, VA), 30 g/L sucrose. Adjust pH to 5.8 with 1*M* KOH, add 10 g/L bactoagar (Difco, Detroit, MI), and autoclave. Pour in Magenta GA7 boxes (Magenta Corporation, Milwaukee, WI), 80 mL/box.
2. Rooting medium: 1X Murashige and Skoog salts and vitamins, 30 g/L sucrose, 0.1 mg/L 1-naphtalene acetic acid (NAA). Adjust pH to 5.8 with 1*M* KOH, and add 8 g/L bacto-agar. Autoclave and cool media to 60°C. Add the appropriate selective agent to select transformed plants. Pour in Magenta GA7 boxes (80 mL/box).
3. K3 medium *(33)*: *See* Table 1 for final concentration of individual components. Prepare stock solutions (*see* Note 1), and add every component of the medium in the order listed. Then adjust to pH 5.8 with 1*M* KOH and autoclave (*see* Note 2).
4. H medium *(34)*: See Table 1 for final concentration of individual components. Prepare stock solution (*see* Note 1); microelement stock solution is the same as for K3. Add every component of the medium in the order listed. Then adjust to pH 5.8 with 1*M* KOH. Filter-sterilize (0.24-μm filter).
5. A medium *(35)*: See Table 1 for final concentration of individual components. Prepare stock solutions (*see* Note 1); the microelement and vitamin stock solutions are the same as for K3. Add every component of the medium in the order listed. Then adjust to pH 5.7 with 1*M* KOH. For liquid medium, add mannitol to a final concentration of 90 g/L (final osmolarity

Table 1
Composition of Media Used to Culture Tobacco Leaf Protoplasts

Component	Concentration			
	K3	A	H	RP
Macroelements (final) concentrations in mg/L[a]				
KNO_3	2500	1010	1900	273
KH_2PO_4		136	170	271
NH_4NO_3	250	800	600	
NaH_2PO_4	150			
$CaCl_2-2H_2O$	900	440	600	
$Ca(NO_3)_2,4H_2O$				416
$Mg(NO_3)_2,6H_2O$				392
$MgSO_4-7H_2O$	250	740	300	57
$(NH_4)_2SO_4$	134			233
NH_4-succinate		50		
Microelements (final) concentration in mg/L[b]				
Na_2EDTA	74.6	74.6	74.6	14.9
$FeCl_3-6H_2O$	27	27	27	5.4
H_3BO_3	3	3	3	1.24
KI	0.75	0.75	0.75	0.16
$MnSO_4-H_2O$	10	10	10	3.38
$ZnSO_4-7H_2O$	2	2	2	1.72
$CuSO_4-5H_2O$	0.025	0.025	0.025	0.005
$NaMoO_4-2H_2O$	0.25	0.25	0.25	0.05
$CoCl_2-6H_2O$	0.025	0.025	0.025	0.005
Vitamins (final) concentration in mg/L[c]				
Myo-inositol	100	100	100	20
Biotin			0.1	
Pyridoxine HCl	1	1	1	0.5
Thiamine HCl	10	10	10	0.1
Nicotinamide			1	
Nicotinic acid	1	1		0.5
Folic acid			0.4	
D-Ca-Pantothenate			1	
p-Aminobenzoic acid			0.02	
Choline chloride			1	
Riboflavin			0.2	
Ascorbic acid			2	
Vitamin A			0.01	

Table 1 *(continued)*

Component	Concentration			
	K3	A	H	RP
Vitamin D$_3$			0.01	
Vitamin B$_{12}$			0.02	
Glycine			0.1	2
Other organic compounds (final concentration in mg/L)				
Caseine hydrolysat			200	
Organic acids mix			d	
Puric/pyrimidic bases mix			e	
Sugars (final concentration in g/L)^f				
Sucrose	102.96	30	0.25	0.5
Xylose	0.25		0.25	
Mannitol		85	0.25	8
Sorbitol			0.25	
Cellobiose			0.25	
Fructose			0.25	
Glucose			68.4	
Mannose			0.25	
Rhamnose			0.25	
Ribose			0.25	
Hormones (final concentration in mg/L)^g				
NAA	1	0.1	1	
2-4D	0.1		0.1	
BAP	0.2	1	0.2	
Zeatin				0.25

[a]Macroelements: make a 10X stock solution of the combined macroelements.

[b]Microelements: Na$_2$ EDTA: make a 200X stock solution. FeCl$_3$-6H$_2$O; make a 200X stock solution. Make a 1000X stock solution of the combined other elements.

[c]Vitamins: Myo-inositol: Add directly to the medium. Make a 100X stock solution of the combined other vitamins.

[d]Citric acid: 40 mg/L, fumaric acid: 40 mg/L, malic acid: 40 mg/L, sodium pyruvate: 20 mg/L. Make a 100X stock solution of the combined elements, and adjust pH to 6.5 with NH$_4$OH.

[e]Adenine: 0.1 mg/L, guanine: 0.03 mg/L, thymine: 0.03 mg/L, uracil: 0.03 mg/L, hypoxanthine: 0.03, cytosine: 0.03 mg/L. Make a 1000X stock solution of the combined elements, and adjust pH to 6.5 with NH$_4$OH.

[f]Sugars: add directly to the medium.

[g]Hormones: Make 100X stock solution in DMSO (Sigma) of each hormone. NAA: α-naphtalene acetic acid; 2-4D: 2,4-Dichlorophenoxyacetic acid; BAP: 6-Benzylaminopurine.

of 620 mosM), and autoclave. For solid medium, 30 g/L mannitol and add 8 g/L bacto-agar. Autoclave and cool media to 60°C. Add the appropriate selective agent to select transformed plants and pour into Petri dishes.

6. RP medium (regeneration) *(13)*: *See* Table 1 for final concentration of individual components. Prepare stock solutions (*see* Notes 1 and 3). Add every component of the medium in the order listed. Then adjust to pH 5.8 with 1*M* KOH. Add 8 g/L bacto-agar, autoclave, and cool media to 60°C. Add the appropriate selective agent to select transformed plants. Pour into Petri dishes.

7. Domestos: 11% solution (Lever, Warrington, UK).

2.3. Protoplast Electroporation Media

1. Leaf digestion medium: K3 medium with 136.9 g/L sucrose (instead of 102.96 g/L), 1.2% cellulase R-10 "Onozuka" (Yakult Honsha Co, LTD, Tokyo, Japan), and 0.8% macerozyme R-10 (Yakult Honsha Co, LTD). Stir the solution for 1 h with a magnetic stirrer in a cold room (8°C). Centrifuge the solution for 10 min at 5000*g*, 5°C, to pellet the insoluble particles that would prevent subsequent filtration. Filter the supernatant through a layer of Whatman #5 paper in the cold room. Adjust to pH 5.8 with 1*M* KOH, and filter-sterilize (0.24-μm filter). Store 20-mL aliquots at –20°C for up to a year.

2. W5 medium: 154 m*M* NaCl, 125 m*M* CaCl$_2$, 5 m*M* KCl, 5 m*M* glucose. Adjust to pH 6 with 1*M* NaOH. Autoclave.

3. Electroporation buffer: 80 m*M* KCl, 4 m*M* CaCl$_2$, 2 m*M* potassium phosphate, pH 7.2, 8% mannitol (final osmolarity is 540 mosM).

2.4. DNA Extractions

DNA is extracted and quantified using standard molecular biology techniques. Genomic DNA is extracted from leaves according to Dellaporta et al. *(36)* and further purified on CsCl gradient *(37)*. Genomic DNA fragments should be about 50 kbp in size, as measured on 0.2% agarose gel *(37)*. Plasmid DNA is prepared by an alkaline lysis method *(37)* (*see* Note 4). Plasmid pHP23 has been used to confer kanamycin resistance *(38)*.

2.5. DNA Solutions

Plasmid DNA, carrier DNA (e.g., sheared herring sperm DNA), and genomic DNA are sterilized by one phenol/chloroform extraction, one chloroform extraction, and ethanol precipitation. Ethanol precipitation alone is not sufficient to ensure sterility. After centrifugation following ethanol precipitation, dry the pellet under a sterile flow bench, and resuspend in sterile distilled water at a concentration of 1 µg/µL for plasmid DNA and genomic DNA, and 2.5 µg/µL for carrier DNA.

3. Methods

3.1. Production of Axenic Tobacco Shoot Culture

Keep an in vitro culture of tobacco plants (*see* Note 5) in GA7 Magenta boxes by subculturing shoots on a regular basis on tobacco shoot culture medium. This culture is started from seeds sterilized by immersion for 10 min in 11% Domestos, followed by five washes in sterile distilled water. When not in use, the culture can be kept in a cold room for several months with an occasional light source. When planning to carry out an electroporation experiment, the tobacco plants should be subcultured 5–6 wk before protoplast isolation.

3.2. Isolation of Tobacco Leaf Protoplasts

1. Pour 10 mL of digestion medium into a 90-mm diameter sterile plastic Petri dish. Use 1 Petri dish/g of tissue, which will produce $2–2.5 \times 10^6$ protoplasts. Each construct or condition tested requires 10^6 protoplasts $(3 \times 3 \times 10^5)$.
2. Cut out a leaf from in vitro tobacco plantlets. Lay it in the bottom of a sterile plastic Petri dish. Excise the central vein with forceps and blade. Superpose 5–6 half-leaves in the bottom of the Petri dish, and cut them transversally into 3–5-mm strips. Lay the strips on the digestion medium with the lower face in contact with the liquid (*see* Note 6). Cover the surface of the 90-mm Petri dish entirely; this represents approx 1 g of leaf tissue. Seal the Petri dish with Nescofilm, and incubate the Petri dishes at 25°C in the dark overnight (16 h) without shaking.
3. After overnight digestion, most of the protoplasts are still in place in the leaf tissue. Gently tap two Petri dishes against each other in such a way that the leaf strips dissociate, freeing the protoplasts (*see* Note 7).
4. Pipet 10 mL from each dish several times to free as many protoplasts as possible. The easiest way to do this is to use an electric pipetter (e.g., Drummond, Broomall, PA).
5. Fit a sterile 100-µm sieve (*see* Note 8) on top of a sterile 50-mL glass beaker. Pipet the protoplast suspension with a sterile 10-mL pipet, and pour it over the mesh of the sieve. Light tapping of the beaker-sieve unit against the bench might be necessary to initiate the flow through the mesh. The contents of up to four Petri dishes can be filtered using the same sieve and beaker.
6. Pour about 10 mL of the protoplast suspension into an 11-mL sterile plastic screw-cap centrifuge tube. To prevent contamination, seal the top of the tube with Nescofilm. Centrifuge at 60*g* for 5 min at room temperature.

7. After the centrifugation, the protoplasts will float on the medium, and the leaf debris will be pelleted. The medium should be slightly green because some protoplasts neither float nor pellet. Transfer the protoplast band using a Pasteur pipet to a new centrifuge tube. Ideally, the volume containing the protoplasts should not exceed 1 mL. Leave a thin layer of protoplasts behind. Pool protoplasts from two tubes. Add 0.4M sucrose and 80 mM KCl to a final volume of 10 mL, replace the top, seal, and centrifuge at 60g for 5 min at room temperature.

8. Repeat step 7.

9. Repeat step 7, but before centrifugation, measure the protoplast density in an aliquot with a hemocytometer and calculate the total number of protoplasts present in the tube.

10. After the centrifugation, resuspend the protoplasts in electroporation buffer at a density of 1.5×10^6/mL, allowing a loss of 0.25×10^6 protoplasts from the total number of protoplasts calculated in step 9. Check the density with the hemocytometer, and then adjust to a density of 1×10^6 protoplasts/mL.

11. Incubate for 2 h at 8°C in the dark (*see* Note 9).

3.3. Electroporation for Transient-Expression Studies

1. Using a Pipetman P1000 and a large-bore tip, transfer 0.3 mL of protoplasts to a 1-mL electroporation cuvet with a 0.4-cm electrode gap (*see* Note 10).

2. Add 10 µg (30 µg/mL final concentration) of the plasmid of interest; no carrier DNA is needed (*see* Note 11).

3. Apply one pulse, 225 V/cm, capacitance 100 µF, and time constant 72 ms (*see* Note 12).

4. Transfer the protoplasts to 11-mL sterile plastic tubes for about 45 min to allow cell membranes to reseal.

5. Dilute each 0.3-mL sample with 3 mL of A medium (620 mosM) containing cefotaxime (100 mg/L final concentration). Incubate protoplasts at 25°C in the dark for 36–40 h. Store the tubes in a nearly horizontal position so that the liquid does not reach the top of the tube. This is easily achieved by placing a Petri dish at the top end of a horizontal tube holder (*see* Note 13).

6. Add 7 mL of W5 buffer to each tube, and centrifuge at 170g for 10 min to pellet the protoplasts. Discard the supernatant, and invert the tube for 1 min to drain the excess liquid.

7. Resuspend the pellet in 200 µL of extraction buffer, and transfer to a 1.5-mL tube. At this stage, the sample can be stored at –80°C, or extracted directly by adding an equal volume of sand and grinding with a disposable pestle (*see* Note 14).

8. Assay transient expression (e.g., ref. *39*).

3.4. Electroporation for Stable-Transformation Studies

1. Using a Pipetman P1000 and a large-bore tip, transfer 0.3 mL of proto-plasts into a 1-mL electroporation cuvet with a 0.4-cm electrode gap (*see* Note 10).
2. Add 10 μg of the plasmid of interest and 50 μg of carrier DNA.
3. Apply two pulses, 225 V/cm, capacitance 100 μF, and time constant 48 ms (*see* Note 15).
4. Transfer the protoplast sample to a 90-mm sterile plastic Petri dish, and incubate at room temperature for 45 min to allow cell membranes to reseal.
5. Autoclave (or melt in a microwave oven if already sterile) 15 mL of K3 medium containing 1.2% Seaplaque low-melting agarose (FMC Bio-Products, Rockland, ME)/1×10^6 electroporated protoplasts. Cool to 80°C, and add an equivalent volume of H medium. Cool to 30°C in a water bath. Add 100 mg/L cefotaxime. Perform step 6 before the medium starts to solidify (approx 2 h).
6. Add 10 mL of liquid K3/H-0.6% agarose to the 3×10^5 protoplasts (0.3 mL electroporated sample) present in each 90-mm Petri dish. Let the agarose set for 1 h, and seal with Nescofilm.
7. Incubate 24 h at 25°C in the dark.
8. Incubate 9 d at 25°C in continuous light or 16 h light + 8 h dark.
9. With a spatula, cut the agarose in each Petri dish into strips 1 cm wide. Then cut each strip transversally in two. Lift each half-strip with a 1-cm wide spatula, and lay them in a Petri dish containing 20 mL of A medium + mannitol, plus selection agent (e.g., kanamycin at 50 mg/L). Transfer each K3/H medium Petri dish into two Petri dishes (1.5×10^5 protoplasts/ Petri dish). Seal with Nescofilm.
10. Incubate dishes with the agarose strips at 25°C, in continuous light or 16 h light + 8 h dark on an orbital shaker at 80 rpm.
11. After 4–5 wk when the calli reach 1–2 mm, transfer to solid A medium plus selection agent.
12. When the calli reach 5 mm in diameter, transfer to RP regeneration medium plus selection agent.
13. When shoots are 1–5 mm long, cut them at their base, and transfer to a GA7 box containing rooting medium plus selection agent. Avoid transfer-ring any callus tissue.
14. Only the plantlets forming roots in the presence of a selective agent should be considered as true transformants. Once the roots are formed, wash the agar from the root system under running tap water, and transfer the plant-lets to soil (50:50 mix of vermiculite and soil in $100 \times 100 \times 100$ mm plastic pots). To aid cuticle development, which is needed to control water losses, cover each transformant with a Magenta GA7 box. Incubate for 3 d,

and then tilt the box for a day before removing it totally. Depending on your particular culture conditions (humidity level), you may need to adjust the period of time required for cuticle development.

15. Water the plants with commercially available nutritive solution, following the specifications of the manufacturer.

3.5. Electroporation
of Genomic DNA into Plant Protoplasts

1. Using a Pipetman P1000 and a large-bore tip, transfer 0.3 mL of protoplasts (*see* Note 10) into an electroporation cuvet with a 0.4-cm electrode gap.
2. Add 10 μg of the plasmid pHP23 and 17 μg of genomic DNA (*see* Note 16).
3. Apply two pulses, 225 V/cm, capacitance 100 μF, and time constant 48 ms (*see* Note 15).
4. Perform steps 4–11 of Section 3.4.
5. When the calli reach 5 mm, they can be subjected to the second screen. The calli are picked with forceps and transferred to solid A medium (without mannitol) supplemented with the second selection agent (e.g., chlorsulfuron at 2 μg/L). If the selection scheme used cannot be applied at the callus stage, apply selection at step 8 (below) on whole plants.
6. Calli that remain green on the second selection medium are transferred to RP regeneration medium supplemented with the second selection agent.
7. Perform steps 13–15 of Section 3.4.
8. If applicable, subject the plants to the selection scheme for your gene of interest.
9. Assess genetic linkage of plasmid and the gene of interest by segregation analysis *(24)*.

4. Notes

1. The different stock solutions are kept at 8°C. If prepared on a regular basis, they do not need to be autoclaved except for the macroelement stock solution. Before use, check for signs of contamination. The osmolarity of each liquid medium can be changed by modifying the mannitol concentration so the protoplasts neither burst nor shrivel.
2. The vitamin stock solution for K3 is available from Duchefa Biochemie (Haarlem, The Netherlands).
3. The vitamin stock solution for RP medium is the same as for MS and is available from Duchefa Biochemie. The microelements in RP medium are at 1/5 the final concentration of the microelements in MS medium.
4. The basal level of expression of inducible promoters can be reduced by using a methylation-deficient *E. coli* strain to amplify the plasmid DNA *(40)*.

5. We use *Nicotiana tabacum* cv "Petit Havana" mutant SR1 *(41)*, because it stays relatively small when cultured in soil compared to other cultivars. In $10 \times 10 \times 10$ cm pots, this cultivar reaches 80–100 cm in height. It also regenerates well in in vitro culture. Other tobacco cultivars can, however, be electroporated with success (e.g., Samsun, Xanthi, W38) using the method described here. Protoplasts prepared from plantlets grown on MS20 *(32)* gave a 25% higher transient expression than protoplasts prepared from plantlets grown on MS30 *(32)*, and protoplasts prepared from tobacco plants grown in the greenhouse had a higher voltage optimum: 300 V/cm (data not shown).

6. We found that leaf tissue digested better when its lower face was in contact with the digestion medium. Depending on your protoplast source, you may need to adjust the osmotic pressure of the K3 medium so the protoplasts neither burst nor shrivel by modifying the mannitol concentration.

7. Depending on the leaf batch used, digestion of the strips should be more or less complete after overnight digestion. At this stage, the protoplasts can be examined under a microscope. They should appear spherical with green chloroplasts inside. A few protoplasts will appear huge and contain no chloroplasts. These will be eliminated during the purification procedure.

8. Sieves can be bought from Sigma. Alternatively, sieves can be made with a 100-µm nylon mesh and a soft plastic 50–100 mL beaker, which should be autoclavable and slightly conical. With a scalpel, cut 3-cm slices of the beaker. Using two slices, block the nylon mesh between the smaller and larger beaker slices, glue the two slices, or heat-seal them by piercing a hole through them with red-hot forceps.

9. Protoplasts that are allowed to settle at 8°C for 2 h survive electrical shock better. If left too long, wall synthesis starts, and the protoplasts become refractory to transfection. Wall synthesis can be temporarily inhibited to improve transient expression *(42)*.

10. Large-bore pipet tips are made by cutting the end of the tips with a scalpel 2–3 mm from its end. To ensure sterility, plug tips with cotton.

11. For transient-expression assays, we prefer to omit carrier DNA to eliminate possible interference with the test plasmid. However, the addition of 150 µg/mL of carrier DNA increases expression by twofold over the level with 30 µg/mL of plasmid alone. Expression levels with 150 µg/mL of carrier and 10 µg/mL of plasmid were equivalent to that obtained with 30 µg/mL of plasmid alone. Most workers use 50 µg/mL carrier, but we found a maximum effect with 150 µg/mL.

12. Depending on your culture conditions and tobacco cultivar used, optimization might be required. We found that leaf protoplasts from in vitro cultured tobacco transfect optimally at 225 V/cm, but tobacco leaf protoplasts

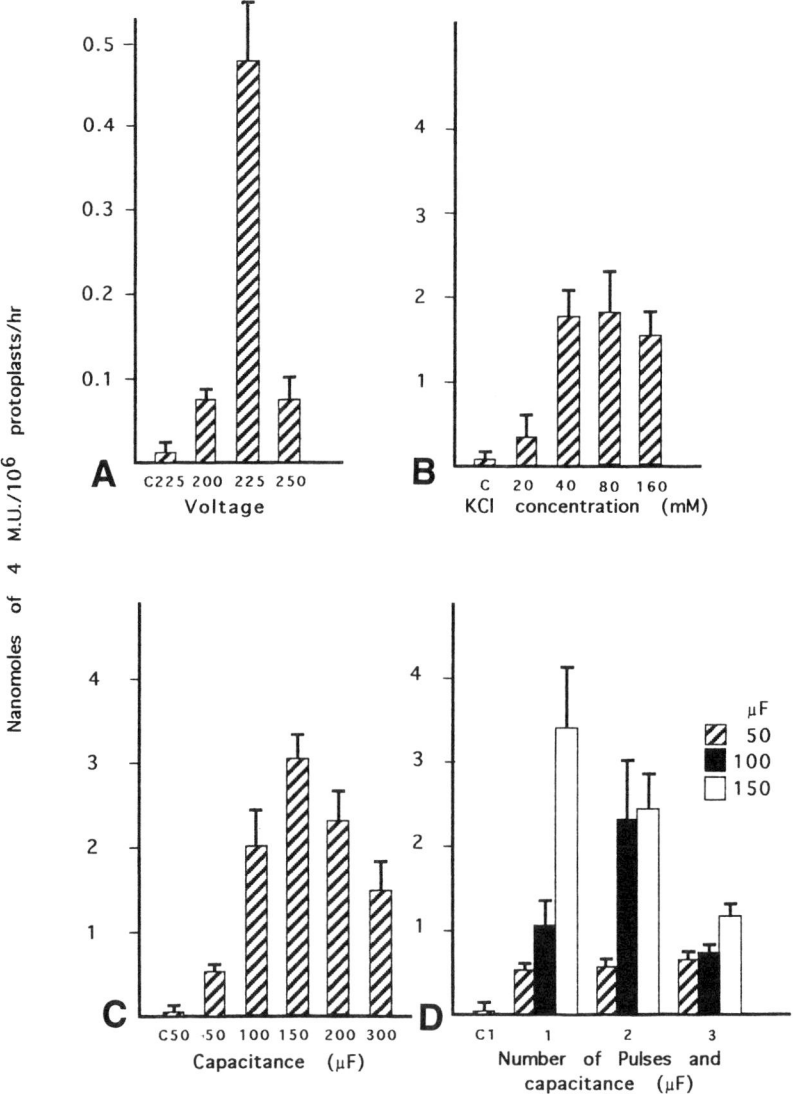

from greenhouse-grown tobacco required 300 V/cm. It is likely that several combinations of parameters will give similar degrees of DNA uptake in a given type of cell. We chose to optimize sequentially the charge and size of the capacitor, the resistance of the circuit, and the number of electrical pulses. We first fixed the capacitor size to 50 μF and the KCl concentration to 20 m*M*. The optimum with our conditions and mesophyll protoplasts prepared from in vitro plantlets was 225 V/cm (Fig. 1A). The pulse duration was then altered by changing the KCl concentration in the electroporation medium. The KCl concentration showed an optimum between 40 and 80 mM (Fig. 1B). A concentration of 80 mM was selected for further optimization, because some protoplast batches showed a clear optimum at 80 mM (data not shown). With this KCl concentration, protoplasts were electroporated using different capacitances ranging between 50 and 300 μF. As a result, the time constant varied from 23–138 ms. A setting of 150 μF (69 ms) gave the highest transient expression (Fig. 1C). Higher capacitance gave no greater β-glucuronidase (GUS) activity and resulted in lower protoplast survival. The optimum pulse number was then determined for three different capacitances (Fig. 1D). For 50 μF, the transient GUS activity increased slightly from one to three pulses (and decreased at five pulses; data not shown). For 100 μF, there was an optimum at two pulses before the activity decreased at three pulses; 150 μF and one pulse gave the highest transient expression, two pulses being detrimental.

13. An alternative to culturing the protoplasts in a horizontal tube is to use 3-cm diameter Petri dishes. However, directly culturing in a tube eliminates a transfer step when protoplasts are collected. Tubes are incubated horizontally to increase the liquid/air interface.

14. An alternative to grinding is sonication.

15. For stable transformation, protoplasts were electroporated in the presence of plasmid pHP23 *(38)*, which confers kanamycin resistance. Transformation frequencies obtained with a combination of various capacitances and pulse number are shown in Table 2. A transformation efficiency of 1.1 ×

Fig. 1. *(previous page)* Effect of different electroporation parameters on transient expression in tobacco protoplasts. Error bars represent the standard deviation of three replicate treatments. pGUS (J. Topping, University of Leicester, UK) consists of a promoterless GUS reporter gene (derived from pBI101 *[42]*). pCGUS (pBI221) *(42)* consists of a 35S CaMV promoter, a GUS coding region, and an NOS terminator inserted into pUC19. **(A)** GUS activity in protoplasts electroporated at 50 μF, 20 mM KCl with pGUS (C) at 225 V/cm, or pCGUS at 200, 225, or 250 V/cm. **(B)** GUS activity in protoplasts electroporated at 50 μF, 225 V/cm with pGUS (C), or pCGUS in increasing KCl concentration: 20, 40, 80, or 160 mM. **(C)** GUS activity in protoplasts electroporated at 225 V/cm, 80 mM KCl with increasing capacitance: 50, 100, 150, 200, or 300 μF. **(D)** GUS activity in protoplasts electroporated at 225 V/cm, 80 mM KCl with different capacitances (50, 100, or 150 μF) and increasing pulse number (1, 2, or 3).

Table 2
Effect of Capacitance Size and Pulse Number on the Transformation Frequency

Capacitance, μF	Number of pulses	Number of kanamycin resistant colonies[a]		Absolute transformation frequency × 10^{-3}
		pHP23	Control	
50	1	330	0	0.33
50	3	554	0	0.55
100	1	501	0	0.50
100	2	1153	0	1.15
100	3	948	0	0.95
150	1	924	0	0.92
150	3	759	0	0.76

[a]The number of kanamycin resistant colonies and absolute transformation frequencies (transformants per initial number of electroporated protoplasts) are given for an initial number of 10^6 protoplasts/experiment. Colonies were counted 6 wk after beginning of selection on kanamycin.

10^{-3} transformants/electroporated protoplast was reached with a capacitance of 100 μF and two pulses. Nearly the same efficiency was obtained with a capacitance of 150 μF and one pulse. Further transformation experiments using 100 μF and two pulses always gave an efficiency of 0.8–1.5 × 10^{-3}.

16. As a negative control, use genomic DNA extracted from a close relative of the donor plant.

Acknowledgments

The authors wish to thank M. G. K. Jones and M. De Both for help and advice, and G. Hull, R. Cooke, and J. A. Nickoloff for editing the manuscript.

References

1. Dron, M., Cloux, S. D., Dixon, R. A., Lawton, M. A., and Lamb, C. J. (1988) Glutathione and fungal elicitor regulation of a plant defense gene promoter in electroporated protoplasts. *Proc. Natl. Acad. Sci. USA* **85,** 6738–6742.
2. Shulze-Lefert, P., Becker, A. M., Shulze, W., Hahlbrock, K., and Dangl, J. (1989) Functional architechture of the light-responsive chalcone synthase promoter from parsley. *Plant Cell* **1,** 707–714.
3. Walker, J. C., Howard, E. S., Dennis, E. S., and Peacock, W. J. (1987) DNA sequences required for anaerobic expression of the maize alcohol dehydrogenase 1 gene. *Proc. Natl. Acad. Sci. USA* **84,** 6624–6628.
4. Marcotte, W. R., Bayley, C. C., and Quatrano, R. S. (1988) Regulation of a wheat promoter by abscissic acid in rice protoplasts. *Nature* **335,** 454–457.

5. Huttly, A. K. and Baulcombe, D. C. (1989) A wheat alpha-amylase2 promoter is regulated by gibberellin in transformed oat aleurone protoplasts. *EMBO J.* **8,** 1907–1913.
6. Hattori, T., Vasil, V., Rosenkrans, L., Hannah, L. C., Mc Carthy, D. R., and Vasil, I. K. (1992) The viviparous-1 gene and abscisic acid activate the C1 regulatory gene for anthocyanin biosynthesis during seed maturation in maize. *Genes Dev.* **6,** 609–618.
7. Wirtz, U., Schell, J., and Czernilofski, A. (1987) Recombination of selectable marker DNA in *Nicotiana tabacum. DNA* **6,** 245–253.
8. Restrepo, M. A., Freed, D. D., and Carrington, J. C. (1990) Nuclear transport of potyviral proteins. *Plant Cell* **2,** 987–998.
9. Carrington, J. C., Freed, D. D., and Leinicke, A. J. (1991) Bipartite signal sequence mediates nuclear translocation of the plant potyviral Nla protein. *Plant Cell* **3,** 953–962.
10. Fujiwara, T., Naito, S., Chino, M., and Nagata, T. (1991) Electroporated protoplasts express seed specific gene promoters. *Plant Cell Reports* **9,** 602–606.
11. Beilmann, A., Pfitzner, A. J., Goodmann, H. M., and Pfitzner, M. (1991) Functional analysis of the pathogenesis-related 1a protein gene minimal promoter region. *Eur. J. Biochem.* **196,** 415–421.
12. Negrutiu, I., Shillito, R., Potrykus, I., Biasini, G., and Sala, F. (1987) Hybrid genes for the analysis of transformation conditions. I—setting up a simple method for direct gene transfer in plant protoplasts. *Plant Mol. Biol.* **8,** 363–373.
13. Installe, P., Negrutiu, I., and Jacobs, M. (1985) Protoplasts-derived plants in *N. plumbaginifolia*: improving the regeneration response of wild type and mutant cultures. *J. Plant Physiol.* **119,** 443–454.
14. Shewry, P. R., Tatham, A., Halford, N. G., Barker, J. H. A., Hannapel, U., Gallois, P., Thomas, M., and Kreis, M. (1994) Opportunities for manipulating the seed protein composition of wheat and barley in order to improve quality. *Transgenic Res.* **3,** 3–12.
15. Gallois, P., Lindsey, K., Malone, R., Kreis, M., and Jones, M. G. K. (1992) Gene rescue in plants by direct gene transfer of total genomic DNA into protoplasts. *Nucleic Acids Res.* **20(15),** 3977–3982.
16. Giraudat, J., Hauge, B. M., Valon, C., Smalle, I., Parcy, F., and Goodman, H. M. (1992) Isolation of the *Arabidopsis ABI3* gene by positional cloning. *Plant Cell* **4,** 1251–1261.
17. Coen, E. S., Romero, J. M., Doyle, S., Elliot, R., Murphy, G., and Carpenter, R. (1990) *Floricaula*: a homeotic gene required for flower development in *Antirrhinum majus. Cell* **63,** 1311–1322.
18. Feldmann, K. A. (1991) T-DNA insertion mutagenesis in *Arabidopsis*: mutational spectrum. *Plant J.* **1-1,** 71–82.
19. Simoens, C., Alliotte, Th., Mendel, R., Muller, A., Schiemann, J., Lijsebettens, M., Schell, J., Van Montagu, M., and Inzé, D. (1986) A binary vector for transferring genomic libraries to plants. *Nucleic Acids Res.* **14,** 8073–8090.
20. Prosen, D. E. and Simpson, R. B.(1987) Transfer of a ten-number genomic library to plants using Agrobacterium tumefaciens. *BioTechnology* **5,** 966–971.

21. Klee, H. J., Hayford, M. B., and Rogers, S. G. (1987) Gene rescue in plants: a model system for "shotgun" cloning by retransformation. *Mol. Gen. Genet.* **210,** 282–287.
22. Olszewski, N. E., Martin, F. B., and Ausubel, F. M. (1988) Specialised binary vector for plant transfomation: expression of the *Arabidopsis thaliana* ALS gene in *Nicotiana tabacum. Nucleic Acid Res.* **16–22,** 10,765–10,782.
23. Lazo, G. R., Stein, P. A., and Ludwig, R. A. (1991) A DNA transformation-competent *Arabidopsis* genomic library in *Agrobacterium. BioTechnol.* **9,** 963–967.
24. Gallois, P., Lindsey, K., Malone, R., Kreis, M., and Jones, M. G. K. (1992) Gene rescue in plants by direct gene transfer of total genomic DNA into protoplasts. *Nucleic Acids Res.* **20,** 3977–3982.
25. Arumuganathan, K. and Earle, E. D. (1991) Nuclear DNA content of some important plant species. *Plant Mol. Biol. Rep.* **9–3,** 208–218.
26. Bennett, M. D., Smith, J. B., and Heslop-Harrison, J. S. (1982) Nuclear DNA amounts in angiosperm. *Proc. R. Soc. Lond.* **B216,** 179–199.
27. Bennett, M. D. and Smith, J. B. (1976) Nuclear DNA amounts in angiosperm. *Phil. Trans. R. Soc. Lond.* **B274,** 227–274.
28. Lowy, I., Pellicer, A., Jackson, J. F., Sim, G. K., Silverstein, S., and Axel, R. (1980) Isolation of transforming DNA: cloning the hamster aprt gene. *Cell* **22,** 817–823.
29. Perucho, M., Hanahan, D., Lipsich, L., and Wigler, M. (1980) Isolation of the chicken thymidine kinase gene by plasmid rescue. *Nature* **285,** 207–210.
30. Goldfarb, M., Shimizu, K., Perucho, M., and Wigler, M. (1982) Isolation and preliminary characterisation of the human transforming gene from T24 bladder carcinoma cells. *Nature* **296,** 404–409.
31. Littman, D. R., Thomas, Y., Maddon, P. J., Chess, L., and Axel, R. (1985) The isolation and sequence of the gene encoding T8: a molecule defining functional classes of T lymphocytes. *Cell* **40,** 237–246.
32. Murashige, T. and Skoog, F. (1962) A revised medium for rapid growth and bioassays with tobacco tissue cultures. *Plant Physiol.* **15,** 473–497.
33. Nagy, J. I. and Maliga, P. (1976) Callus induction and plant regeneration from mesophyll protoplasts of *N. Sylvestris. Z. Pflanzenphysiol.* **78,** 453–455.
34. Kao, K. N., Constabel, F., Michayluk, R., and Gamborg, O. L. (1974) Plant protoplast fusion and growth on intergeneric hybrid cells. *Planta* **120,** 215–217.
35. Caboche, M. (1986) Nutritional requirement of protoplasts derived, haploid tobacco cells grown at low densities in liquid medium. *Planta* **149,** 7–18.
36. Dellaporta, S. L., Wood, J., and Hicks, J. B. (1983) A plant DNA minipreparation: version II. *Plant Mol. Biol. Rep.* **1,** 19–21.
37. Sambrook, J., Fritsch, E. F., and Maniatis, T. (1989) *Molecular Cloning: A Laboratory Manual.* Cold Spring Harbor Laboratory, Cold Spring Harbor, NY, p. 545.
38. Paszkowski, J., Baur, M., Bogucki, A., and Potrykus, I. (1988) Gene targeting in plants. *EMBO J.* **7,** 4021–4026.
39. Jefferson, R. A., Kavangh, T. A., and Bevan, M. W. (1987) GUS: β-glucuronidase as a sensitive and versatile gene fusion marker in higher plants. *EMBO J.* **6,** 3901–3907.

40. Torres, J. T., Block, A., Hahlbrock, K., and Somssich, I. E. (1993) Influence of bacterial strain genotype on transient expression of plasmid DNA in plant protoplasts. *Plant J.* **4(3),** 587–592.

41. Etzold, T., Fritz, C. C., Shell, J., and Schreier, P. H. (1987) A point mutation in the chloroplast 16S rRNA gene of a streptomycin resistant *Nicotiana tabacum. FEBS Lett.* **219,** 343–346.

42. Chapel, M. and Glimelius, K. (1990) Temporary inhibition of the cell wall synthesis improves the transient expression of the GUS gene in *Brassica napus* mesophyll protoplasts. *Plant Cell Rep.* **9,** 105–108.

CHAPTER 8

Electroporation of *Brassica*

Frank Siegemund and Klaus Eimert

1. Introduction

A main objective of electroporation is to transform cells by stable incorporation and expression of foreign genetic material. Incorporation may result in transformation with marker genes, as well as desired traits from a breeder's point of view (e.g., male sterility, virus resistance, insect resistance). By using electroporation, a variety of plant species refractory to other transformation techniques have been transformed.

The Brassicaceae includes many important vegetables. Breeding efforts are very intensive, especially in regard to the use of cytoplasmic male sterility for hybrid seed production. Because of the importance of these crop plants, the availability of efficient gene-transfer techniques is highly valuable. Some reports concerning genetic transformation of members of the genus *Brassica,* mainly *B. napus,* were published. There are fewer reports of transformation of *Brassica oleracea* by *Agrobacterium*-mediated gene transfer and microinjection (C. Beclin, F. Charlot, C. Dove, personal communication; A. Mukhopadhyay, A. Pradhan, D. Pental, personal communication; *1*). However, from work with *B. napus,* it is known that the efficiency of regeneration from different primary explants is tremendously decreased after infection with *Agrobacterium tumefaciens* and following decontamination *(2,3).* This led researchers to establish procedures for direct gene transfer in *Brassica.*

The use of electroporation has been reported for *B. napus (4–7*; P. Bergman and K. Glimelius, personal communication) and *B. oleracea* var. *botrytis (8).* This chapter describes the transformation of *B. oleracea*

From: *Methods in Molecular Biology, Vol. 55: Plant Cell Electroporation and Electrofusion Protocols* Edited by: J. A. Nickoloff Humana Press Inc., Totowa, NJ

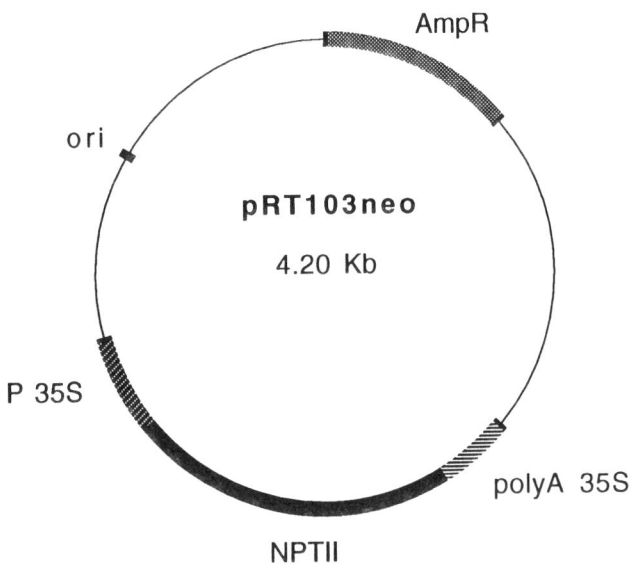

Fig. 1. Structure of plasmid pRT103neo. The NPTII gene expression is regulated by the cauliflower mosaic virus 35 promoter (P 35S) and polyadenylation (polyA) signal sequences.

var. *botrytis* (cauliflower) by means of electroporation of protoplasts as well as the analysis of transformed calli and regenerated plants.

2. Materials
2.1. Plasmid Vectors

Plasmids pRT103neo and pRT99gus, both constructed by R. Töpfer (Cologne), are suitable for gene-transfer experiments. pRT99gus (6.71 kbp) contains the neomycin phosphotransferase gene (NPTII) under control of the CaMV35S promotor and terminator, as well as the coding sequence of the β-glucuronidase gene (GUS) from *E. coli,* also under control of the same regulatory elements *(9,10).* This construct is particularly useful for the investigation of transient gene expression. pRT103neo (4.2 kbp) contains the kanamycin resistance conferring NPTII gene (Fig. 1).

2.2. Solutions for Culture and Staining of Protoplasts

1. CPW solution: 20 mM KH_2PO_4, 1 mM KNO_3, 10 mM $CaCl_2$, 1 mM $MgSO_4$, 1 μM KI, 1 μM $CuSO_4$, pH 5.8.

2. Enzyme solution E1 *(15)*: 1% Cellulase Onozuka R-10 (Serva, Heidelberg, Germany), 0.5% macerozyme OnozukaR-10 (Serva), 1X CPW solution, and 9% mannitol, pH 5.8. Sterilize the enzyme solution for protoplast isolation by passing through a 0.22-μm membrane filter. Store aliquots at –20°C.
3. W5 protoplast washing solution: 150 mM NaCl, 125 mM CaCl$_2$ · 6H$_2$O, 5 mM KCl, 5 mM glucose, pH 5.8.
4. Protoplast cultivation medium: components of H1, H18, H20, H22, PP3, and MSO are given in Table 1.
5. Modified Murashige-Skoog medium (MSO; *11*): 18.8 mM KNO$_3$, 1.25 mM KH$_2$PO$_4$, 20.6 mM NH$_4$NO$_3$, 1.5 mM MgSO$_4$, 100 μM H$_3$BO$_3$, 100 μM MnSO$_4$, 100 μM FeSO$_4$, 100 μM Na$_2$-EDTA, 0.15 μM CoCl$_2$, 0.1 μM CuSO$_4$, 30 μM ZnSO$_4$, 1 μM Na$_2$MoO$_4$, 5 μM KI, 26.7 μM glycine, 0.56 mM inositol, 4.1 μM nicotinic acid, 2.5 μM pyridoxine-HCl, 0.29 μM thiamine-HCl, 87.8 mM sucrose, pH 5.8. Store at 4°C.
6. Cell-wall stain: Dilute 0.1 mL saturated Calcofluor White ST solution (American Cyanamide Corp., Bound Brook, NJ) with 10 mL of CPW solution containing 13% mannitol.
7. 0.6M sucrose.
8. 0.3M CaCl$_2$.

2.3. Electroporation Buffers

1. MEMg *(16)*: 0.2 μM MgCl$_2$, 5 μM methylethylsulfonate (MES), 0.5M mannitol, pH 5.7.
2. MKCl *(17)*: 5 mM KCl, 22 μM MES, 0.38M mannitol, pH 7.2. Store both solutions at 4°C.

2.4. DNA Preparation

DNA extraction buffer *(19)*: 50 mM Tris-HCl, pH 7.6, 50 mM EDTA. 100 mM NaCl, 10 mM β-mercaptoethanol, 2% sodium-dodecylsulfate (SDS, w/v). Store al 4°C.

2.5. Drug Selection

Filter-sterilize stock solutions (10 mg/mL) of kanamycin or hygromycin, and store in aliquots at –20°C (*see* Note 3).

2.6. NPTII Assays

1. Extraction buffer B: 10% glycerol, 62.5 mM Tris-HCl, pH 6.8, 50 mM DTT, 1% sodium-desoxycholate.
2. 5X Reaction buffer: 335 mM Tris-HCl, 210 mM MgCl$_2$, 2M NH$_4$Cl. Adjust to pH 7.1 using 1M maleic acid.

Table 1
Composition (in mg/L) of the Protoplast Cultivation Media

	H1	H18	H20	H22	PP3	MSO
KNO_3	950	1900	1900	2500	2500	1900
$CaCl_2 \cdot 2H_2O$	440	440	440	150	150	440
$MgSO_4 \cdot 7H_2O$	185	370	370	250	250	370
$(NH_4)_2SO_4$				134	134	
NH_4NO_3	200	1650	1650			1650
$NaH_2PO_4 \cdot H_2O$				150	150	
KH_2PO_4	85	170	170			170
KI	0.41	0.83	0.83	0.75	0.75	0.83
H_3BO_3	3.10	6.20	6.20	3	3	6.2
$MnSO_4 \cdot 4H_2O$	11.1	22.3	22.3	10	10	22.3
$ZnSO_4 \cdot 7H_2O$	4.3	8.6	8.6	2	2	8.6
$Na_2MnO_4 \cdot 2H_2O$	0.25	0.25	0.25	0.25	0.25	0.25
$CuSO_4 \cdot 5H_2O$	0.025	0.025	0.025	0.025	0.025	0.025
$CoCl_2 \cdot 6H_2O$	0.012	0.025	0.025	0.025	0.025	0.025
Na_2EDTA	18.6	37.25	37.25	37.25	37.25	37.25
$FeSO_4 \cdot 7H_2O$	13.9	27.83	27.83	27.83	27.83	27.83
Inositol	100	100	100	100	100	100
Pyridoxine HCl	1	1	0.5	1	1	0.5
Thiamine HCl	10	10	0.1	10	10	0.1
Nicotinic acid	1	1	0.5	1	1	0.5
Kinetin	0.7					
2,4-D[a]	2				0.5	
NAA	2	0.02		0.2	0.2	
BAP	0.3	2	1	1	0.2	
GA_3			0.04			
Glycine						2
Mannitol	70,000					
Glucose	20,000			2500		
Sucrose		30,000	30,000	20,000	30,000	30,000
Ribose				125		
Casein-hydrolysate		150	150	150	150	
pH	5.8	5.8	5.8	5.8	5.8	5.8

[a]Abbreviations used: 2,4-D: 2,4-dichlorphenoxiacetic acid, NAA: nicotinic acetic acid, BAP: benzylaminopurine, GA_3: gibberellic acid.

3. Assay mix: 0.2 vol 5X reaction buffer, 0.01 mM ATP, 0.015 mM kanamycin, 10 mM NaF, 30 µCi $^{32}P(\gamma)ATP$. Extraction buffer and assay mix should be freshly prepared.
4. Phosphate buffer: 14.8 mL 0.066M KH_2PO_4, 85.2 mL 0.066M Na_2HPO_4, pH 7.5.

5. P81 Phosphocellulose paper (Gibco-BRL, Gaithersburg, MD).
6. 10M sodium pyrophosphate, 11 mg/mL ATP.

2.7. Dot-Blot Hybridization

1. Denaturation solution: 0.5M NaOH, 1.5M NaCl.
2. Neutralization solution: 0.5M Tris-HCl, pH 7.4, 1.5M NaCl.
3. 20X SSC buffer: 3M NaCl, 0.3M sodium citrate, pH 7.0.

3. Methods

3.1. Axenic Cultivation of Plants and Isolation of Protoplasts

1. Sterilize the cauliflower seeds by treating with 70% ethanol for 5 min, followed by 3% sodium-hypochlorite for 15 min (*see* Note 1).
2. Wash the seeds thoroughly with sterile water about four times.
3. Place the seeds for germination on MSO medium under illumination for periods of 16 h light, 8 h dark at 28°C.
4. For axenic culture, root the shoot tips and maintain on MSO medium supplemented with 0.2 mg/L nicotinic acetic acid (NAA), and subculture every 2–3 wk.
5. For the isolation of protoplasts, use the fully developed leaves.
6. Slice the surface of the leaves slightly at the lower side, and incubate the leaves in enzyme solution E1 for 14–16 h at 28°C in the dark. Use Petri dishes of about 10 cm diameter. Gentle shaking (25 rpm) increases the yield of protoplasts considerably (*see* Note 2).

3.2. Preparation of Protoplasts for Electroporation and Cultivation

1. Monitor for complete digestion of the plant cell wall by staining with Calcofluor White ST.
2. Mix one drop of the protoplast solution with one drop of cell-wall stain solution on a glass slide.
3. Incubate for 5–10 min at room temperature.
4. Examine under an UV microscope. Cell-wall material will fluoresce yellow.
5. Pass the digested leaf material through a nylon sieve (mesh size 45 μm), and centrifuge for 10 min at 100g to collect protoplasts.
6. Resuspend the protoplasts in 5 mL of 0.6M sucrose solution. Overlay this suspension carefully with 1 mL of 0.3M CaCl$_2$.
7. Centrifuge the gradient for 10 min at 100g. Intact protoplasts will float to the interface between sucrose and CaCl$_2$.
8. Wash these protoplasts twice with W5 solution *(12),* resuspend in electroporation buffer, count using a hemocytometer, and electroporate.

A   **B**   **C**   **D**

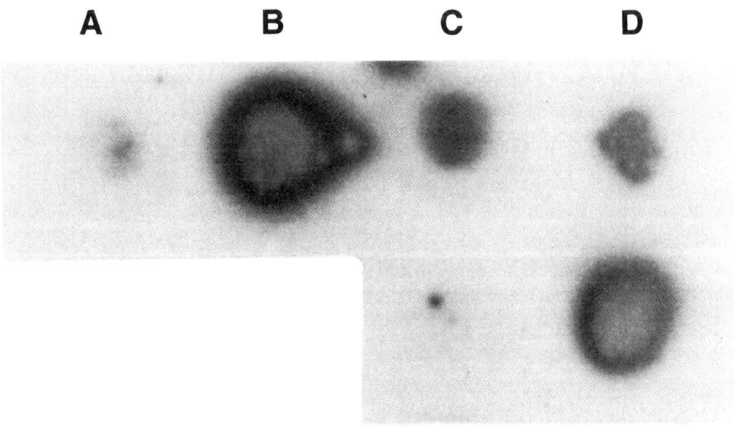

Fig. 2. Transient NPTII expression after direct gene transfer of pRT103neo into cauliflower protoplasts. Top row: **(A)** PEG treatment; **(B)** PEG treatment combined with electroporation; **(C)** electroporation at room temperature; **(D)** electroporation at 0°C. Bottom row: (C) untreated *Brassica* protoplasts; (D) NPTII positive *N. tabacum* SR1 control. Reprinted by permission of Kluwer Academic Publishers.

3.3. Electroporation Procedure

1. Add 10 µg of plasmid DNA and 50 µg of carrier DNA to 10^6 protoplasts resuspended in 1 mL electroporation buffer. Transfer to an electroporation cuvet.
2. Pulse one or more times with 20 µs square wave at 1 kV/cm (*see* Notes 4–7; Fig. 2).
3. After a reconstitution period of 10 min following pulse treatment, wash with W5 protoplast washing solution, and centrifuge the protoplasts.

3.4. Cultivation of Protoplasts

1. Cultivate the protoplasts for the first 2–3 d in H1 medium (modified K3-medium [13], Table 1) in the dark. Optimal plating density is 4×10^5 protoplasts/mL.
2. After 3–5 d (depending on the extent of cell wall regeneration), add 0.5 vol of H22 medium, and continue the cultivation in weak light (400–500 lx) for 3–4 d.
3. Dilute the cultivation medium a second time with 0.5 vol of H22.
4. After a total of 14 d, sediment the regenerated cells and embed into PP3-medium with 0.8% LMP-agarose (Table 1).

5. Cultivate according to Shillito et al. (*14*; "agarose bead type culture"). Cut the protoplasts containing agarose into pieces, and place in liquid medium in a Petri dish. To select for stable transformants, add 50 mg/L kanamycin (Serva, research-grade) or 10 mg/L hygromycin (Serva, research-grade) to the liquid medium of the agarose-bead-type culture.
6. Transfer the emerging microcalli on solid H18 and, after beginning differentiation, on H20 medium.

3.5. Analysis of Transformed Protoplasts

Analysis of transformed protoplasts can be done by examination of the NPTII activity, dot-blot hybridization of genomic DNA, or by Southern hybridization (*see* Note 8).

3.5.1. NPTII Assay

NPTII activity in putative transformants can be tested by dot-blot assay according to McDonnell et al. *(18)*.

1. Homogenize the plant material (e.g., leaf material) in liquid nitrogen after addition of 1 vol of extraction buffer B.
2. Centrifuge for 5 min at 10,000*g*, 4°C.
3. Mix aliquots of 15 μL supernatant and 15 μL assay mix at 7°C, and incubate for 30 min.
4. Blot 20 μL of this mixture on P81 phosphocellulose paper (Gibco-BRL).
5. Soak P81 phosphocellulose paper with 10*M* sodium pyrophosphate, 11 mg/mL ATP, then air-dry.
6. Wash the dry filter for 2 min in hot (80°C) phosphate buffer.
7. Rinse the filter three to five times in phosphate buffer at room temperature, and dry again.
8. Expose the filters to X-ray film for 3–7 d at –70°C or 7–14 d at –18°C.

3.5.2. DNA Isolation

Genomic DNA of protoplasts and calli can be easily extracted and purified as described by Junghans and Metzlaff *(19)*.

1. Grind the plant material in a mortar in liquid nitrogen, and incubate for 1 h in 1–2 vol of extraction buffer at 4°C.
2. Extract the lysate with 0.5 vol of phenol, and centrifuge for 10 min at 3000*g*.
3. To the upper phase, add 0.5 vol of phenol/chloroform and 0.5 vol of chloroform/isoamyl alcohol (24:1). Extract and centrifuge as above.
4. Overlay the upper phase with 1 vol of isopropanol, and carefully invert the centrifuge tube.

5. The settled DNA can be collected with a pipet tip, dried, and resuspended in an appropriate volume of TE buffer.

This procedure is useful for isolating DNA from 10 mg of plant material. Sufficient DNA is recovered from 10^5 protoplasts to detect single-copy genes in a dot-blot assay. Hybridization can be performed with a nick-translated radiolabeled Tn5 probe according to standard procedures *(20)*.

3.5.3. Dot-Blot Hybridization

For dot-blot hybridization, the protocol of Raeder and Broda *(21)* can be used:

1. Mix DNA solution with denaturation solution (1:1), and blot on a nitrocellulose filter.
2. Air-dry for 20 min.
3. Float the filters in neutralization solution for 20 min.
4. Wash the filters in 6X SSC for 5 min.
5. Incubate at 80°C for 2 h before using for hybridization.

Southern hybridization can be done according the standard procedure described by Maniatis et al. *(20)*. Cleavage of the DNA should be done as recommended by the suppliers of the enzymes (*see* Note 9; Figs. 3 and 4).

4. Notes

1. Protoplasts can also be isolated from leaf mesophyll of greenhouse-cultured plants. The sterilization procedure is the same, but the hazard of contamination with microorganisms is much greater. Sometimes commercially available household cleaners (Domestos or Chlorox) are used for disinfection, but we do not recommend their use. The viability of the protoplasts seems to be reduced in comparison to sodium hypochlorite-sterilized material.
2. The yield of viable protoplasts is significantly higher if plants are kept in the dark for 24 h prior to the isolation procedure.
3. Hygromycin is toxic and must be handled very carefully with attention to appropriate safety standards. Using lower concentrations of antibiotics or starting selection later will result in more putative resistant transformants. However, this will increase the number of protoplasts that "escape" from the selection as well.
4. Transient expression of the introduced NPTII gene is detectable in protoplasts following electroporation. When electroporation was used in com-

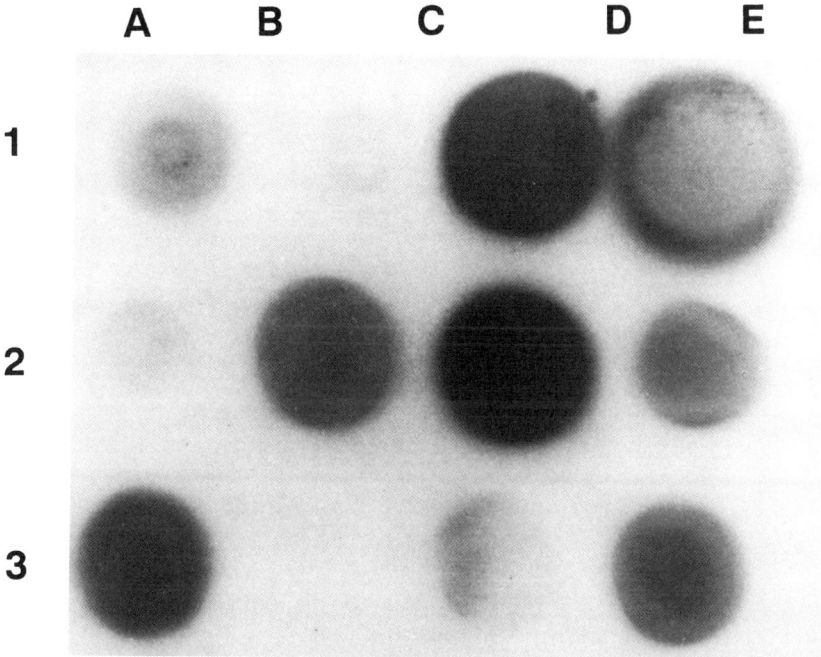

Fig. 3. Dot-blot of genomic DNA of calli obtained after electroporation of protoplasts. Positions A3 and B3 are positive and negative controls, respectively.

bination with PEG, no increase in transient expression was observed respective of the temperature used. However, electroporation without PEG at 4°C resulted in a large increase in transient expression (Fig. 2).

5. The protocol described was used to recover stable transformants of cauliflower following electroporation of a NPTII coding sequence under the control of the CaMV 35S promotor and terminator. Maximal uptake of the foreign DNA was achieved under the following conditions: MKCl buffer, 20-µs pulses, 1 kV/cm field strength. Large amounts of DNA were taken up and expressed by the protoplasts (Fig. 2). The external membrane-bound DNA was digested by DNase treatment prior to DNA isolation and, therefore, did not contribute to the hybridization signal. After application of nine pulses of 160 V/cm with 8.1-ms intervals and 4-µF capacitance, we obtained kanamycin-resistant calli growing on selective medium. The transformation frequency was about 3.5×10^{-5}.

6. Varying the pulse length with PEG-treated cells did not result in a increase of transformation frequency. This result, and the known toxicity of PEG, suggests that these conditions are not optimal.

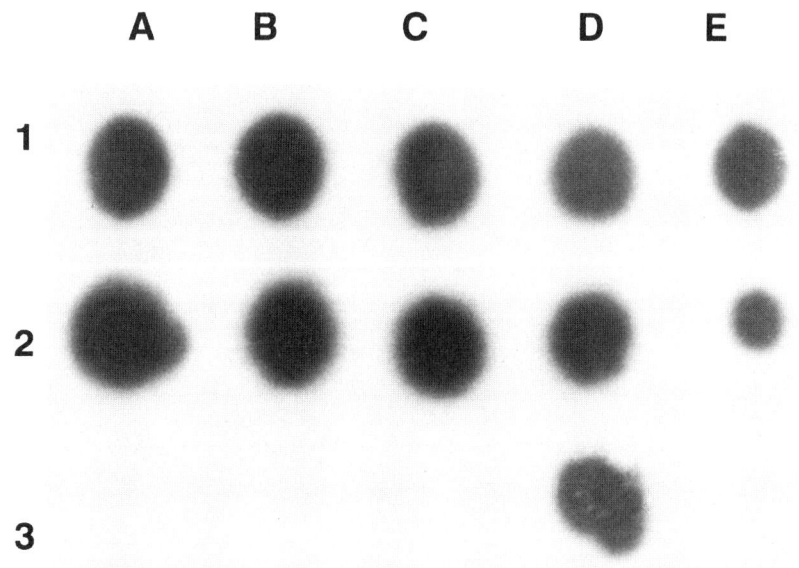

Fig. 4. Expression of the NPTII sequence 3 mo after electroporation. Each signal represents an independent transformant.

7. A stimulating effect of heat-shock treatment prior to electroporation has been described *(22,23)*. We observed this effect only if the protoplasts were cooled to room temperature using ice immediately after heat shock. The transformation frequency may be increased by cooling the protoplasts during electroporation. A possible explanation for this phenomenon is that the low temperature delays resealing of the induced membrane pores.

8. Although DNA uptake into protoplasts after electroporation is distinctly higher than in the PEG-induced transformation, this did not result in a higher frequency of stable transformation. Linearization of the plasmids had no effect on the transformation rates in our experiments.

9. The presence of the NPTII gene could be demonstrated by dot-blot hybridization of genomic DNA with a radioactive labeled Tn5 probe (Fig. 3). The gene was also expressed 3 mo after electroporation in calli grown under selective as well as nonselective conditions (Fig. 4). Although we could demonstrate the incorporation of the NPTII sequence by Southern hybridization, the regeneration of transgenic calli into plants still remains problematic. Shoots could be regenerated only in few cases. Problems concerning the expression of selectable markers in *Brassica* and the regeneration of resistant shoots have been reported by other groups as well *(2,3,24)*. The selection using the bar gene seems to be more reliable *(24)*, so those

planning to transform *Brassica* only for the introduction of a marker gene might consider using this gene. Furthermore, we observed that electroporation treatment significantly increases the plating efficency of treated *Brassica* protoplasts.

References

1. Eimert, K., Schröder, C., and Siegemund, F. (1992) Expression of the NPTII-sequence in cauliflower after injection of agrobacteria into seeds. *J. Plant Physiol.* **140**, 37–40.
2. Charest, P. J., Holbrook, L. A., Gabard, J., Iyer, V. N., and Miki, B. L. (1988) Agrobacterium mediated transformation of thin layer explants from *Brassica napus* L. *Theor. Appl. Genet.* **75**, 438–445.
3. Peecham, P. M. (1988) Successful cocultivation of *Brassica napus* microspores and proembryos with *Agrobacterium. Plant Cell Rep.* **8**, 387–390.
4. Bergman, P. and Glimelius, K. (1993) Electroporation of rapeseed protoplasts transient and stable transformation. *Physiol. Plant.* **88**, 604–611.
5. Guerche, P., De Almeida, E. R. P., Schwarztein, M. A., Cander, E., Krebbers, E., and Pelletier, G. (1990) Expression of the S2 albumin from *Bertholletia excelsa* in *Brassica napus. Mol. Gen. Genet.* **221**, 306–314.
6. Herve, C., Rouan, D., Guerche, P., Montane, M.-H., and Yot, P. (1993) Molecular analysis of transgenic rapeseed plants obtained by direct transfer of two separate plasmids containing, respectively, the cauliflower mosaic virus coat protein and a selectable marker gene. *Plant Sci.* **91**, 181–193.
7. Boyer, J.-C., Zaccomer, B., and Haenni, A.-L. (1993) Electrotransfection of turnip yellow mosaic virus RNA into *Brassica* leaf protoplasts and detection of viral RNA products with non-radioactive probe. *J. Gen. Virol.* **74**, 1911–1917.
8. Eimert, K. and Siegemund, F. (1992) Transformation of cauliflower (*Brassica oleracea* L. var. *botrytis*) an experimental survey. *Plant Mol. Biol.* **19**, 485–490.
9. Jefferson, R. A. (1987) Assaying chimeric genes in plants: the GUS gene fusion system. *Plant Mol. Biol. Rep.* **5**, 387–405.
10. Jefferson, R. A., Kavanagh, T. A., and Bevan, M. W. (1987) GUS fusions: β-glucuronidase as a sensitive and versatile gene fusion marker in higher plants. *EMBO J.* **6**, 3901–3907.
11. Murashige, T. and Skoog, F. (1962) A revised medium for rapid growth and bioassay with tobacco tissue culture. *Physiol. Plant.* **15**, 473–479.
12. Menczel, L. and Wolfe, K. (1984) High frequency of fusion induced in freely suspended protoplast mixtures by polyethylene glycol and dimethylsulfoxide at high pH. *Plant Cell Rep.* **3**, 196–198.
13. Kao, K. N., Constabel, F., Michayluk, M. R., and Gamborg, O. L. (1974) Plant protoplast fusion and growth of intergeneric hybrid cells. *Planta* **120**, 215–227.
14. Shillito, R. D., Paszkowski, J., and Potrykus, I. (1983) Agarose plating and a bead type culture technique enable and stimulate development of protoplast-derived colonies in a number of plant species. *Plant Cell. Rep.* **2**, 244–247.
15. Siegemund, F. and Tschuch, G. (1984) Elektrisch induzierte Fusion von *Lycopersicon*-Protoplasten unterschiedlicher Herkunft. *Arch. Züchtungsforsch.* **14**, 163–168.

16. Negrutiu, I., Shillito, R., Potrykus, I., Biasini, G., and Sala, F. (1987) Hybrid genes in the analysis of transformation conditions. I. Setting up a simple method for direct gene transfer in plant protoplasts. *Plant Mol. Biol.* **8,** 363–373.

17. Guerche, P., Charbonnier, M., Jounanin, J., Tourneur, C., Paszkowski, J., and Pelletier, G. (1987) Direct gene transfer by electroporation in *Brassica napus. Plant Sci.* **52,** 111–116.

18. McDonnell, R. E., Clark, R. D., Smith, W. A., and Hinchee, M. A. (1987) A simplified method for detection of neomycin phosphotransferaseII activity in transformed plant tissues. *Plant Mol. Biol. Rep.* **5,** 380–386.

19. Junghans, H. and Metzlaff, M. (1990) A simple and rapid method for the preparation of total plant DNA. *BioTechniques* **8,** 176.

20. Sambrook, J., Fritsch, E. F., and Maniatis, T. (eds.) (1982) *Molecular Cloning, A Laboratory Manual.* Cold Spring Harbor Laboratory, Cold Spring Harbor, NY.

21. Raeder, U. and Broda, P. (1984) Comparisons of the lignin-degrading white rot fungi *Phanerochaete orysosporum* and *Sprototrichum pulverutentum* at the DNA level. *Curr. Genet.* **8,** 499–506.

22. Shillito, R. D., Saul, M. W., Paszkowski, J., Müller, M., and Potrykus, I. (1985) High efficiency direct gene transfer to plants. *BioTechnology* **3,** 1099–1103.

23. Rhodes, C. A., Pierce, D. A., Mettler, I. J., Mascarenhas, D., and Detmer, J. J. (1988) Genetically transformed maize plants from protoplasts. *Science* **240,** 204–207.

24. De Block, M., De Brouwer, D. D., and Tenning, P. (1989) Transformation of *Brassica napus* and *Brassica oleracea* using *Agrobacterium tumefaciens* and the expression of the bar and neo genes in the transgenic plants. *Plant Physiol.* **91,** 694–701.

CHAPTER 9

Transformation of Maize by Electroporation of Embryos

Carol A. Rhodes, Kathleen A. Marrs, and Lynn E. Murry

1. Introduction

Plant cell walls present a barrier to DNA uptake not present in mammalian cells or plant protoplasts. Transformation methods that force DNA through this network of cellulose fibers include particle gun technology (ballistics; *1*), imbibition of dried seeds *(2),* and use of silicon carbide "whiskers" *(3).* Each of these methods has associated limitations. Ballistics, the bombarding of cells or tissues with high-velocity, DNA-coated microprojectiles, has been used most widely and successfully with both cell cultures *(4–6)* and embryos serving as target tissues. Electroporation has several advantages over ballistics in that it does not require the expensive particle gun apparatus, associated consumable supplies, and licensing. We describe in this chapter an electroporation protocol that has been used successfully to introduce DNA into maize embryos and has resulted in the recovery of stably transformed, fertile, transgenic maize plants.

The development of protocols to insert DNA directly into plant cells, without prior removal of cell walls, greatly expands the genetic manipulations possible with plants. Transformed embryos can be cultured to give rise to mature plants, and their progeny analyzed for inheritance of recombinant DNA. Other plant tissues, including type II calli *(7)* or suspension cell cultures *(8),* may be transformation targets using electropo-

From: *Methods in Molecular Biology, Vol. 55: Plant Cell Electroporation and Electrofusion Protocols* Edited by: J. A. Nickoloff Humana Press Inc., Totowa, NJ

ration if regeneration is not an issue. Particularly if fertile transformants are not needed, the electroporated cell lines may be proliferated and assayed to yield valuable information on plasmid DNA longevity, site(s) of integration, and transient vs stable expression.

Electroporation causes DNA uptake through both the cell wall and the cell membrane. The pulse of electric current may be carrying DNA through interstices in cell walls by electrophoresis, and simultaneously, producing temporary pores in the cell membrane that permit DNA to pass into the cytoplasm. Results from experiments that varied the orientation of target tissues and/or electrode positions relative to the plasmid DNA *(9)* support this theory. Additional support is provided by the production of transgenic maize via low-voltage current, directional electrophoresis of DNA into immature embryos *(10)*, and the general observation that longer electroporation pulse times yield higher numbers of transformed cells, when other factors are equal *(9,11,12)*.

A major advantage of using embryos as a source of recipient cells for DNA transformation is the relative ease and speed with which fertile, transgenic plants can be recovered. Nonchimeric, transgenic plants can be growing in the greenhouse only 12 wk after electroporation of embryos *(11)*. Perhaps the lack of fertility problems associated with these regenerated, transgenic plants may be attributed to the short period the tissue spent in culture. Sterility has been a common problem in studies that used older, established cell cultures as DNA recipients *(5,6,13)*.

The protocol for electroporation of maize embryos described below is based on methods reported by D'Halluin et al. *(11)*, with modifications described by Klöti et al. *(9)* and Songstad et al. *(12)*. As is the case with any successful transformation system, this protocol requires:

1. The acquisition of genes and development of DNA constructs for selectable or detectable markers;
2. Assays for these genes; and
3. An efficient culture system for propagation and regeneration of plant material.

These will be discussed in the following sections as appropriate.

2. Materials
2.1. Plasmid Vectors, Genes, and Their Detection

Table 1 lists genes that may be used in constructs to test or optimize an electroporation protocol. These genes may be interchanged

Table 1
Marker Genes, Plasmids, Substrates, and Assays

Marker gene	Plasmid	Selection/substrate	Assay
β-glucuronidase (GUS)	pZO1016 *(15)*	X-gluc (Clontech)	Histochemical *(20)*
Lc and C1	pBC17 *(18)*	—	Visual, vacuole
Neomycin phospho- transferase (NPT)	pZO921	Kanamycin (Sigma)	Dot blot *(15)*
Phospinotricin acetyl- transferase (PAT)	pDPG165 *(5)*	Basta, bialophos	Spraying *(5)*

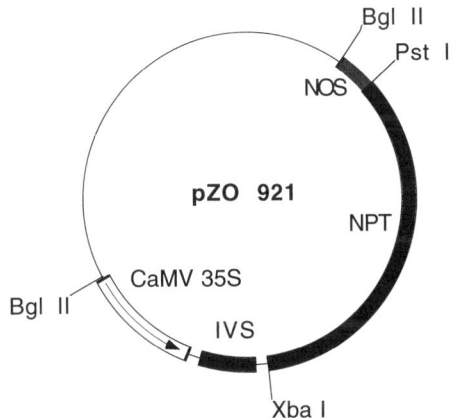

Fig. 1. Diagram of the general structure and components of pZO921. This plasmid is based in a pUC19 vector. The monocot selection cassette contains the 35S promoter from cauliflower mosaic virus, an alcohol dehydrogenase intron (IVS), the neomycin phosphotransferase (NPT) from *E. coli,* and a nopaline synthase (NOS) terminator region.

in a basic plasmid cassette that has been designed for monocot expression. One example is pZO921 (Fig. 1), which contains a 35S promoter from cauliflower mosaic virus, an alcohol dehydrogenase intron (IVS; *14*), the neomycin phosphotransferase (NPT) gene, and a nopaline synthase (NOS) terminator. By cloning vector-compatible polylinker sequences around each new gene of interest, the NPT gene in this plasmid may be replaced with other selectable or detectable markers, such as phosphinotricin acetyltransferase (PAT) or β-glucuronidase (GUS), respectively.

Selection for stable transformants requires use of dominant genes conferring resistance to some growth-inhibiting agent, usually antibiotics or herbicides (e.g., NPT and PAT, respectively). Various assays may be used to confirm the presence (PCR; *15*) or expression of these introduced genes (NPT; *16*, PAT; *17*). Transgenic cells can also be identified by their expression of visible markers, such as Lc and C1. These two genes trigger anthocyanin production in most maize genotypes *(18)*. GUS *(19,20)* is a particularly useful gene for both transformation and expression experiments in that it can be detected histochemically, and quantified spectrophotometrically or fluorimetrically *(21)*.

2.2. Embryos

Any maize genotypes that readily form callus capable of plant regeneration on in vitro culture should respond well in this procedure (*see* Note 1). Genotypes of two public inbreds, H99 and Pa91, and the germ plasm known as HiType II (derived from maize genotypes A188 and B73; *see* Note 2) have been transformed using this protocol. The best embryos will come from vigorously growing maize plants that have been self- or sib-pollinated or outcrossed to another genotype. If grown under greenhouse conditions, daylight should be supplemented with 16 h of light from sodium halide and mercury vapor lamps. Temperatures should be approx 25°C during the daytime and 20°C at night. Embryos are excised from ears about 10–13 d postpollination as described in Section 3.

2.3. Solutions and Reagents for Electroporation

Use double-distilled deionized water to prepare all media and solutions. Sigma Chemicals (St. Louis, MO) makes a complete line of plant cell-culture reagents for the solutions listed below, except where indicated.

1. 10% Sodium hypochlorite.
2. Enzyme solution: 0.3% macerozyme (Kinki Yakult Mfg. Co., Nishinomiya, Japan), 10% mannitol, in a solution of $0.6M$ mannitol, $0.002M$ dithiothreitol, 3 mM MES, pH 5.6 *(22)*. Sterilize with a 0.22-μm filter, and store at 4°C for up to 4 d or freeze at –20°C.
3. N6aph: N6 salts (Gibco-BRL [Gaithersburg, MD]; cf. *23*) with 6 mM asparagine, 12 mM proline, 1 mg/L thiamine-HCl, 0.5 mg/L nicotinic acid, 100 mg/L casein hydrolysate, 100 mg/L inositol, 30 g/L sucrose, 54 g/L mannitol *(24)*. Autoclave or filter-sterilize, and store at 4°C for use within 1–2 mo. For solid media, add 1.6 g/L Phytagel and autoclave.

4. Electroporation media (EPM): 80 mM KCl, 5 mM CaCl$_2$, 10 mM HEPES, 0.425M mannitol, pH 7.2. Store at 4°C up to several months.
5. EPM with 0.2 mM spermidine: Add 290 mg spermidine to 100 mL EPM and filter-sterilize. Prepare fresh the day of electroporation.
6. N6aph plus 2,4-D: supplement N6aph with 1 mg/L 2,4-dichlorophenoxy-acetic acid (2,4-D).

2.4. Solutions and Media for Selection and Recovery of Transformed Plants

1. Kanamycin stock solution: 50 mg/mL, store at –20°C for up to 6 mo. Add filter-sterilized kanamycin to autoclaved N6 medium that has been cooled to 55°C.
2. Media A: N6aph supplemented with 0.2M mannitol, 1.6 g/L Phytagel, and 200 mg/L kanamycin sulfate.
3. Media B: N6aph supplemented with 1.6 g/L Phytagel and 200 mg/L kanamycin.
4. Media C: N6aph with 6 mg/L benzyl-aminopurine (BAP).
5. Media D: Murashige and Skoog (MS; *25*) with 150 mg/L asparagine, 1.0 mg/L thiamine-HCl, and 6% sucrose.
6. Media E: MS with 2% sucrose.

Store autoclaved media at 4°C, and use within 1 or 2 mo. Be sure the media is well sealed to avoid dehydration during storage.

3. Methods

3.1. Preparation of Maize Embryos and Electroporation

1. Remove husks from pollinated ears, and sterilize the entire ear in 10% sodium hypochlorite solution for 20 min; rinse ear three times in sterile water.
2. Remove embryos from ear aseptically in a laminar air flow hood. Carefully slice off the top half of each kernel of a row (or several rows) with a fresh razor blade. Insert a sterile spatula down into the endosperm and under the embryo. Remove the embryo with a brief, twisting motion of the spatula. (Practice this technique with older ears, ~25 d postpollination, which have larger embryos, and work back down to very small embryos.) Note that sterility is absolutely essential for recovery of stable transformants and also improves the reliability of transient assays by eliminating microbial contamination, which can interfere with assays of marker genes.
3. Excise late stage 2 or early stage 3 embryos *(26,27)* from kernels when the embryos are 1.0–1.5 mm long, usually 10–13 d postpollination, under good growing conditions.

4. Treat isolated embryos for 1–3 min in enzyme solution (*see* Note 3). Do not incubate for more than 3 min, since this may cause a decrease in frequency of callus initiation and vigor. Wash embryos thoroughly (at least three rinses, up to 1 h/rinse) with N6aph to remove digestive enzymes and nucleases.

5. After washing, place between 5 and 150 embryos into an electroporation chamber, either a cuvet or other small vessel fitted with electrodes (*see* Notes 4–10). Add EPM or EPM with 0.2 mM spermidine and 10–20 µg plasmid DNA to a total volume of 200 µL.

6. Incubate at room temperature for 1 h, and then transfer the cuvet to an ice bath for 10 min (*see* Note 11). Apply a single electrical pulse of 375 V/cm from a 900-µF capacitor. Immediately after electroporation, add an additional 200–400 µL of N6aph medium to the cuvet, and transfer it back to ice for 10 min.

7. Place embryos with embryo axis downward onto growth medium N6aph plus 2,4-D to stimulate scutellar cells to form callus.

3.2. Selection of Transformed Cells

1. Place embryo axis downward on agarose-solidified growth medium to which a selective agent has been added. Only cells that have integrated a functional copy of the selectable marker gene will proliferate. If NPT is the selectable gene and kanamycin is the selective agent, include 200 mg/L kanamycin in the growth medium (Media A).

2. Incubate plates in the dark at 23°C. Transfer embryos or calli to fresh selective medium (Media B) every 14 d.

3. As the tissues grow, transfer only that which is most vigorous and white or yellow in color. Generally, stable transformants are obvious 4–6 wk after electroporation (*see* Note 12).

3.3. Regeneration of Transformed Plants

After approx 6–8 wk, callus may be transferred to regeneration medium. Several media sequences have been used successfully to recover plants. One example is as follows:

1. Incubate tissue in the dark on Media C for 1 wk.
2. Transfer to Media D, and incubate tissue in the dark for 1 wk.
3. Transfer to Media E, and incubate in the light for 2 wk.
4. Transfer green shoots to larger containers with MS, no hormones, and 1.5% sucrose; place the container under bright lights (90 µE/m/s) for best root development.
5. When roots are sufficient, remove plant from container. Carefully rinse all media from roots before placing plant in moistened potting soil. Keep the transplant in high humidity for 2–4 d.

6. Grow plant to maturity under light conditions simulating strong daylight, such as those described for growing embryo-donor plants in the greenhouse.

4. Notes

1. Donor plants that are growing vigorously are essential for obtaining healthy embryos that will form callus readily. If the necessary light and temperature conditions described in Section 3. cannot be provided, then ears from plants grown elsewhere, in greenhouses or under field conditions, can be utilized. Ears with intact husks should be wrapped tightly in plastic film to prevent dehydration and stressing of embryos, placed over ice in a styrofoam container, and shipped by overnight delivery. After shipment, ears can be sterilized and handled the same as freshly harvested ears. Much of the variation in experimental results is attributable to embryo differences among ears; even donor plants grown under identical conditions can show considerable developmental variation. If the size of an experiment requires the use of embryos from more than one ear, the embryos from multiple ears should be distributed over all treatments in a blocked experimental design.

2. According to Songstad et al. *(12)*, culturing isolated embryos on N6 medium with 1 mg/L 2,4-D, 100 mg/L casamino acids, 25 mM proline, and 10 µM AgNO3 *(24)* for 4 d at 23°C in darkness produced threefold more transformed cells/embryo than electroporating freshly isolated embryos. This result was obtained with the HiType II genotype and may vary for genotypes that respond differently to in vitro conditions. This pretreatment is worth investigating when working with other genotypes.

3. Some reports indicate that enzymatic wounding of cells with protoplasting enzymes before transformation is necessary or increases frequency *(10,11)*. Others *(9,12)* show that no pretreatment is required. Although Songstad et al. *(12)* reported good transformation rates in the absence of enzyme pretreatment, genotype and other factors were not identical to those used by D'Halluin et al. *(11)*. Klöti et al. *(9)* found no difference between transformation rates of electroporated wheat embryos either with or without 0.3% macerozyme pretreatment. Our results with corn embryos indicated predigestion is not necessary, but transformation frequency increased in some cell types or tissues after pretreatment. Because species, genotype, and target tissues were not identical in these studies, it is difficult to extrapolate the necessity or usefulness of this step. It is an obvious candidate for future empirical work.

4. Bio-Rad (Richmond, CA) makes a Gene Pulser model with capacitance extender that produces exponential decay of the electrical discharge from capacitors. Alternatively, electroporation equipment can be constructed

from purchased components following the specifications in Fromm et al. *(28)*. Electroporation chambers are disposable cuvets into which stainless-steel electrodes may be inserted (4-mm gap between electrodes). Other configurations of electroporation chambers are possible and, perhaps, desirable. For example, a custom-made round chamber with accommodations for precise positioning of embryos was used quite successfully with wheat embryos *(9)*. If a chamber with different gap lengths between electrodes is used, voltage must be adjusted to maintain the same magnitude (V/cm) of electric field.

5. Electroporation buffer and voltage conditions should be optimized because these variables affect both pulse length and cell viability. To measure the effects of electroporation on tissue viability, plate out a subsample of embryos on N6aph *(24)* medium and watch for normal callus initiation. Viability must be examined relative to transformation frequency in order to recover the highest number of fertile, stable transformants.

6. Depending on whether the supply of embryos or DNA is more limiting, the number of embryos per cuvet and/or the amount of DNA per cuvet can be altered. Songstad et al. *(12)* recommended 5 embryos and 20 µg plasmid DNA/cuvet (total volume of 200 µL) for high transformation rates and low variability. Klöti et al. *(9)*, however, found a linear increase in transformed cells with increasing amounts of DNA, up to 100 µg/mL, when using 10 embryos/cuvet.

7. The use of circular vs linear plasmid DNA appears to be a matter of choice. Klein et al. *(4)* and Gordon-Kamm et al. *(5)* used linear DNA; Klöti et al. *(9)* and Murry et al. *(15)* used circular plasmid. Since these studies did not produce data that were directly comparable, elucidation of the most efficient form of the DNA awaits accumulation of comparable data on effective gene integration and stable inheritance.

8. The electrolyte used in the electroporation buffer may affect transformation frequency as well as subsequent viability of the embryos. Songstad et al. *(12)* found that $0.07M$ sodium glutamate was better for maize embryos than were buffers containing chloride ions at higher or lower concentrations. The electroporation buffer used by Klöti et al. *(9)* for wheat embryos also avoided chloride ions while including aspartic and glutamic acids as electrolytes. This buffer contains 35 mM aspartic acid, 35 mM glutamic acid, 5 mM calcium gluconate, 5 mM MES, and $0.4M$ mannitol.

9. Any change in electrolyte concentration will affect optimum field strength and pulse length. Generally, longer pulse lengths give higher transformation rates unless the concentration of the electrolyte is either too high, and therefore toxic to the tissues, or too low to move the DNA efficiently through the cell wall and membrane(s).

10. Klöti et al. *(9)* showed that orientation of embryos with respect to the positive and negative electrodes can affect the transformation rate. With oriented embryos, they reported a sevenfold increase in the number of cells expressing the transgene. As discussed in Section 1., a substantial portion of DNA movement into cells may be the result of electrophoresis of DNA (a negatively charged molecule) toward the positive electrode. To permit orientation of embryos, Klöti and coworkers used a custom-designed chamber in which embryos were positioned on a thin layer of agarose facing the negative electrode. Other electroporation chambers, cuvet or otherwise, that enable exact placement of embryos may also enhance transformation rates.

11. Incubation of the embryos on ice for 10 min before and after electroporation may not yield the most efficient transformation rates. Songstad et al. *(12)* found that incubation of embryos at 37°C for 10 min gave fourfold higher transformation rates than incubation on ice and nearly twofold higher than room temperature, 22°C.

12. As with other transformation methods, differences in electroporation efficiency among genotypes may require some optimization of the basic protocol. Anyone applying this electroporation procedure should test variations of the protocol with their tissue, genotype, and species of interest. The best procedure may not be the one that yields the highest number of transformed cells per embryo, but the one that produces the most efficient recovery rate of stable transformants. In order to recover stable transformants, transgenic cells must be capable of cell division and of surviving the chosen selection system.

13. Subtle differences in materials or procedures may make it difficult to reproduce the results cited in this chapter. Using the GUS marker, Klöti et al. *(9)* found that a higher proportion of scutellar cells were transformed than were cells of the shoot of wheat embryos; rice embryos produced opposite results. Although Songstad et al. *(12)* were not able to reproduce the leaf sheath results of Dekeyser et al. *(29),* they successfully developed an electroporation procedure that worked well with corn embryos.

14. Both the GUS and anthocyanin marker genes work well with maize embryos. Expression of these genes can give an estimate of the number of target cells that are transformed and remain viable. Such estimates are essential to the recovery of transformed cell lines and plants. The anthocyanin system requires the correct maize genotype that is present in most commercial maize hybrids. Expression of these two markers may appear to differ slightly, since anthocyanin synthesis occurs in vacuoles and GUS is a cytoplasmic enzyme.

15. Clonally propagated tissue or parts of regenerated plants can be tested for expression of genes that were inserted. GUS expression can be assayed in leaf or root tissue as described by Jefferson et al. *(21)*. PAT activity in leaves can be detected by applying spots of Basta to leaves *(5)*. To detect NPT, a 2% solution of kanamycin containing 0.2% SDS may be applied to an incision made near the midvein of a leaf on 4-wk-old plants. After 8–10 d, susceptible plants will exhibit bleached leaves, whereas resistant plants remain green *(11)*.

References

1. Klein, R. M., Wolf, E. D., and Sanford, J. C. (1987) High velocity microprojectiles for delivering nucleic acids into living cells. *Bio/Technology* **24,** 384–386.
2. Senaratna, T., McKersie, B. D., Kasha, K. J., and Procunier, J. D. (1991) Direct DNA uptake during the imbibition of dry cells. *Plant Sci.* **79,** 223–228.
3. Kaeppler, H. F., Somers, D. A., and Rines, H. (1992) Silicon carbide fiber-mediated stable transformation of plant cells. *Theor. Appl. Genet.* **84,** 560–566.
4. Klein, T. M., Kornstein, L., Sanford, J. C., and Fromm, M. E. (1989) Genetic transformation of maize cells by particle bombardment. *Plant Physiol.* **91,** 440–444.
5. Gordon-Kamm, W. J., Spencer, T. M., Mangano, M. L., Adams, T. R., Daines, R. J., Start, W. G., O'Brien, J. V., Chambers, S. A., Adams, W. R., Willetts, N. G., Rice, T. B., Mackey, C. J., Krueger, R. J., Kausch, A. P., and Lemaux, P. G. (1990) Transformation of maize cells and regeneration of fertile transgenic plants. *Plant Cell* **2,** 603–618.
6. Fromm, M. E., Morrish, F., Armstrong, C., Williams, R., Thomas, J., and Klein, T. M. (1990) Inheritance and expression of chimeric genes in the progeny of transgenic maize plants. *Bio/Technology* **8,** 833–839.
7. Armstrong, C. L. and Green, C. E. (1985) Establishment and maintenance of friable, embryogenic maize callus and the involvement of L-proline. *Planta* **164,** 207–214.
8. Krzyzek, R. A., Laursen, C. R. M., and Anderson, P. C. (1991) Stable transformation of maize cells by electroporation. PCT/US91/09619.
9. Klöti, A., Iglesias, V. A., Wunn, J., Burkhardt, P. K., Datta, J. S. K., and Potrykus, I. (1993) Gene transfer by electroporation into intact scutellum cells of wheat embryos. *Plant Cell Rep.* **12,** 671–675.
10. Murry, L. E., Dietrich, P. S., Pleu, S. H., Lawrence, J. L., and Sinibaldi, R. M. (1994) Transformation in maize through low voltage electric current, in *Biotechnology in Agriculture and Forestry* vol. 25 (Bajaj, Y. S. P., ed.), Springer Verlag, Berlin, pp. 252–261.
11. D'Halluin, K., Bonne, E., Bossut, M., De Beuckeleer, and Leeman, J. (1992) Transgenic maize plants by tissue electroporation. *Plant Cell* **4,** 1495–1505.
12. Songstad, D. D., Halaka, F. G., DeBoer, D. L., Armstrong, C. L., Hinchee, M. A. W., Ford-Santino, C. G., Brown, S. M., Fromm, M. E., and Horsch, R. B. (1993) Transient expression of GUS and anthocyanin constructs in intact maize immature embryos following electroporation. *Plant Cell, Tissue, Organ Cul.* **33,** 195–201.
13. Walters, D. A., Vetsch, C. S., Potts, D. E., and Lundquist, R. C. (1992) Transformation and inheritance of a hygromycin phosphotransferase gene in maize plants. *Plant Mol. Biol.* **18,** 189–200.

14. Mascarenhas, D., Mettler, I. J., Pierce, D. A., and Lowe, H. W. (1990) Intron-mediated enhancement of heterologous gene expression in maize. *Plant Mol. Biol.* **15,** 913–920.

15. Murry, L. E., Elliott, L. G., Capitant, S. A., West, J. A., Hanson, K. K., Scarafia, L., Johnston, S., DeLuca-Flaherty, C., Nichols, S., Cunanan, D., Dietrich, P. S., Mettler, I. J., Dewald, S., Warnick, D., Rhodes, C., Sinibaldi, R. M., and Brunke, K. (1993) Transgenic corn plants expressing MDMV strain B coat protein are resistant to infections of maize dwarf mosaic virus and maize chlorotic mottle virus. *Bio/Technology* **11,** 1559–1564.

16. McDonnell, R. E., Clark, R. D., Smith, W. A., and Hinchee, M. A. (1987) A simplified method for the detection of neomycin phosphotransferase II activity in transformed plant tissue. *Plant Mol. Biol. Rep.* **5,** 380–386.

17. Donn, G., Knipe, B., Malvoisin, P., and Eckes, P. (1990) Field evaluation of glufosinate tolerant crops bearing a modified PPT-acetyltransferase gene from *Streptomyces virido-chromogenes. J. Cell Biochem.* **14E,** 298.

18. Goff, S. A., Klein, T. M., Roth, B. A., Fromm, M. E., Cone, K. C., Radicella, J. P., and Chandler, V. L. (1990) Transactivation of anthocyanin biosynthetic genes following transfer of B regulatory genes into maize tissues. *EMBO J.* **9,** 2517–2522.

19. Oard, J. H., Paige, D. F., Simmonds, J. A., and Gradziel, T. M. (1990) Transient gene expression in maize, rice, and wheat cells using an airgun apparatus. *Plant Physiol.* **92,** 334–339.

20. Jefferson, R. A., Kavanagh, T. A., and Bevan, M. W. (1987) GUS fusions: β-glucuronidase as a sensitive and versatile gene fusion marker in higher plants. *EMBO J.* **6,** 3901–3907.

21. Jefferson, R. A. (1987) Assaying chimeric genes in plants: the GUS fusion system. *Plant Mol. Biol. Rep.* **5,** 387–405.

22. Frearson, E. M., Power, J. B., and Cocking, E. C. (1973) The isolation, culture, and regeneration of petunia leaf protoplasts. *Dev. Biol.* **33,** 130–137.

23. Chu, C. C., Wang, C. C., Sun, C. S., Hsu, C., Yin, K. C., Chu, C. Y., and Bi, F. Y. (1975) Establishment of an efficient medium for anther culture of rice through comparison experiments on the nitrogen sources. *Sci. Sinica* **18,** 659–668.

24. Songstad, D. D., Armstrong, C. L., and Peterson, W. L. (1991) AgNO3 increases type II callus production from immature embryos of maize inbred B73 and its derivatives. *Plant Cell Rep.* **9,** 699–702.

25. Murashige, T. and Skoog, F. (1962) A revised medium for rapid growth and bioassays with tobacco tissue cultures. *Physiol. Plant.* **15,** 473–497.

26. Abbe, E. C. and Stein, O. L. (1954) The growth of the shoot apex in maize: embryogeny. *Am. J. Bot.* **41,** 285–293.

27. Poethig, R. S., Coe, E. H., and Johri, M. M. (1986) Cell lineage patterns in maize embryogenesis: a clonal analysis. *Dev. Biol.* **117,** 392–404.

28. Fromm, M., Taylor, L. P., and Walbot, V. (1985) Expression of genes transferred into monocot and dicot plant cells by electroporation. *Proc. Natl. Acad. Sci. USA* **82,** 5824–5828.

29. Dekeyser, R. A., Claes, B., De Rycke, R. M. U., Habets, M. E., Van Montague, M. C., and Caplan, A. B. (1990) Transient gene expression in intact and organized rice tissues. *Plant Cell* **2,** 591–602.

CHAPTER 10

Transient Gene Expression Analysis in Electroporated Maize Protoplasts

Kathleen A. Marrs and J. C. Carle Urioste

1. Introduction

Direct DNA transformation is currently the method of choice for obtaining transformed cereals, such as *Zea mays,* owing to the inherent host limitations of *Agrobacterium tumefaciens.* This soil bacterium, which naturally acts by transferring its own genes into a host plant, is commonly engineered for transferring foreign genes into plants, but its host range is limited mainly to dicots and noncereal monocots, and generally cannot be used to transfer foreign genes into maize *(1,2).* Among the methods for direct (*Agrobacterium*-independent) gene transfer, electroporation has proven to be an extremely efficient and general mechanism for introducing DNA, RNA, and other macromolecules into maize protoplasts (cells in which the cell wall is enzymatically removed prior to electroporation).

DNA entering maize protoplasts via electroporation is able to move to the nucleus, be transcribed, and either exist extrachromosomally or stably integrate into the host genome *(3).* Although the aim of the many transformation experiments is to obtain stably transformed transgenic plants, direct DNA uptake by plant cells has also proven extremely useful as a means for analyzing rapidly the expression and regulation of specific introduced genes via transient-expression analysis. Transient assays can be used as a first step to determine the usefulness of a particular DNA construct prior to the labor-intensive process of generating

From: *Methods in Molecular Biology, Vol. 55: Plant Cell Electroporation and Electrofusion Protocols* Edited by: J. A. Nickoloff Humana Press Inc., Totowa, NJ

maize plants stably transformed with that construct. In addition, transient gene expression assays also provide a very rapid means to characterize the regulation of a gene and expression of its RNA and protein to determine its regulation by various stimuli, since the introduced genes are frequently expressed and regulated in a manner reflecting the natural expression of the genes.

Information about the expression and regulation of genes can also be gathered by fusing the regulatory regions of a gene of interest with a reporter gene having an easily assayed product, such as firefly luciferase (LUC) or *E. coli* β-glucuronidase (GUS). Both LUC and GUS produce enzymes that are easily assayed within hours of transformation *(4–6)*, a hallmark of a good reporter gene. Reporter genes can be driven by regulatory regions of a gene of interest, such as a promoter or enhancer, or their expression modified by different 5' leaders, 3' untranslated regions, or introns. In this way, transient assays have been used to define regulatory regions within plant promoters responsible for heat inducibility *(7)*, ABA responsiveness *(8)*, and anthocyanin biosynthesis *(9)*, among many other examples, as well as to define features of introns important for RNA splicing *(10)*. Questions regarding mRNA stability or RNA turnover have also been addressed using transient expression assays with reporter gene constructs containing 5' and 3' untranslated regions of a gene of interest with the fate of the mRNA or protein produced being followed *(11,12)*. In addition, transient-expression assays using in vitro synthesized mRNA have been developed to study, for example, the role of poly (A) tail length, a feature that cannot be controlled otherwise *(12,13)*.

We describe here the use of transient-expression analysis in electroporated maize Black Mexican Sweet (BMS) protoplasts to monitor gene expression using luciferase as a reporter gene. The protocol we describe here can also be used for the electroporation of maize callus tissue and, with minor modifications as described *(6,14,15)*, can be adapted to rice, carrot, tobacco, and bean suspension cultures.

2. Materials
2.1. Equipment

1. A sterile laminar flow hood, preferably dedicated to plant tissue culture, is necessary both for the routine maintenance of cultures and during electroporation.
2. An electroporation device for protoplast electroporation (*see* Note 1).

3. A luminometer to measure LUC activity (*see* Note 1) or, using the improved LUC assay reagents we describe in this chapter, a standard lab scintillation counter.
4. A fluorimeter to measure GUS activity (*see* Note 1).

2.2. Maize Suspension Cell Cultures

1. Sterile 125-mL Erlenmeyer flasks and foam plugs dedicated for tissue-culture use only.
2. Maize BMS suspension culture line (*see* Note 2).

2.3. Media and Solutions for Cell-Culture Maintenance, Protoplast Preparation, and Electroporation

Use double-distilled, deionized water to prepare all media and solutions. Sigma Chemicals (St. Louis, MO) makes a complete line of plant cell-culture reagents for the solutions listed below, except where indicated.

1. 1000X vitamin stock: Dissolve 130 mg nicotinic acid, 25 mg thiamine, 25 mg pyroxidine, and 25 mg pantothenic acid in 100 mL water. Store 1-mL aliquots at –20°C.
2. 2,4-Dichlorophenoxyacetic acid (2,4-D), 1 mg/mL stock: Solubilize 10 mg 2,4-D in a small volume of ethanol, and adjust volume to 10 mL with water. Store at 4°C for up to 2 mo.
3. BMS growth medium (MS-B) per liter: 1 package Murashige and Skoog (M&S) salts (Gibco, Grand Island, NY), 1 mL 1000X vitamins, 130 mg asparagine, 20 g sucrose, 200 mg myo-inositol, 2 mL 2,4-D stock. Adjust to pH 5.8 with concentrated KOH, autoclave or filter-sterilize through a 0.45-μm Nalgene filter (Nalge, Rochester, NY), and store at room temperature for up to 2 mo (*see* Note 3).
4. Conditioned MS-B medium: Centrifuge and recover the supernatant from the BMS culture(s) to be electroporated. Sterilize through a 0.45-μm Nalgene filter on the day of electroporation for use in PGM (*see* point 8).
5. Protoplast isolation medium (PIM) per liter: 45 g mannitol, 7.35 g CaCl$_2$ · 2H$_2$O, 0.82 g anhydrous sodium acetate. Adjust to pH 5.8 with concentrated HCl, autoclave or filter-sterilize through a 0.45-μm Nalgene filter, and store at room temperature (will keep indefinitely).
6. PIM + enzymes: To 100 mL PIM add 0.3 g CELF cellulase (Worthington, Freehold, NJ), 1.0 g cytolase (Galactomannonase; Genecor, South San Francisco, CA), 20 mg Y23 pectolyase (Seishin Pharmaceuticals, Tokyo, Japan), 0.5 g bovine serum albumin (BSA), and 50 μL 2-mercaptoethanol. Stir to dissolve. Centrifuge at 3000*g* for 5 min to pellet insoluble material, and filter-sterilize through a 0.45-μm Nalgene filter. The final solution is a

clear dark brown in color. Store 50-mL aliquots at −20°C. The solution can be refrozen after thawing, but enzyme activity will be reduced (*see* Note 4).

7. Poration solution (POR) per liter: 36.4 g mannitol, 8.95 g KCl, 0.58 g NaCl, 0.59 g CaCl$_2$, 2.38 g HEPES. Adjust to pH 7.2 with concentrated KOH. Autoclave or filter-sterilize through a 0.45-μm Nalgene filter, and store at room temperature (will keep indefinitely).

8. Protoplast growth medium (PGM): Prepare MS-B medium as described above, but bring the final volume to 800 mL only and adjust to pH 5.8 with concentrated KOH. Add 54.6 g mannitol, autoclave, and store at room temperature for up to 2 mo (*see* Note 3). On day of electroporation, add 1 vol of conditioned medium to 4 vol of PGM (volume needed determined by number of electroporations to be performed).

2.4. Solutions and Reagents for Reporter Gene Expression Analysis

1. LUC extraction buffer (equivalent to Promega Biotech [Madison, WI] CCLR cell-culture lysis reagent): 100 mM potassium phosphate, pH 7.8, 1 mM EDTA, 10% glycerol, 1% Triton X-100, 250 μL 2-mercaptoethanol. Sterilize through a 0.45-μm Nalgene filter, and store at 4°C (will keep indefinitely).

2. Luciferin 5 mM stock: 16 mg luciferin (Luciferin 4-monooxygenase; Analytical Luminescence Laboratories, San Diego, CA) in 10 mL water. Store in the dark at −20°C.

3. LUC assay reagent (LAR): 20 mM Tricine, pH 7.8, 5 mM MgCl$_2$, 0.1 mM EDTA, 3.3 mM DTT, 270 μM coenzyme A (COA), 500 μM luciferin, 500 μM ATP. Store small aliquots in the dark at −20°C.

4. GUS buffer: 50 mM Na$_2$HPO$_4$, pH 7.0, 10 mM 2-mercaptoethanol, 10 mM EDTA, 0.1% sarkosyl, 0.1% Triton X-100 (will keep indefinitely).

5. MUG, 5 mM stock: 1.76 mg/mL Methylumbelliferyl β-D-glucuronide (MUG) in 10 mL GUS buffer. Make fresh each time; 100 μL are needed for each assay.

6. Stop buffer: 0.2M Na$_2$CO$_3$ in water.

7. MU standard, 1 mM stock: 9.9 mg Methyl umbelliferone (MU) in 50 mL water. Dilute 1:10,000 with stop buffer to 100 nM to standardize the fluorimeter.

3. Methods

3.1. Electroporation of Reporter Genes

Reporter genes in flexible cloning cassettes are commercially available for firefly LUC (available from Promega Biotech), and GUS (available from Clontech, Palo Alto, CA). These vectors provide convenient multicloning sites for cloning into all three reading frames, as well as control promoters conferring constitutive expression. The transient gene expression assay we describe here utilizes the simultaneous electroporation of both a LUC reporter gene plasmid as well as a GUS reference plasmid. The use of the internal GUS control plasmid allows reliable comparison between different LUC constructs, by correcting for differences in protoplast viability and electroporation efficiency. Examples of control reporter gene constructs routinely used in our laboratory are shown in Fig. 1.

1. Grow BMS cultures at 25°C in 30 mL of MS-B medium in 125-mL flasks on a shaker platform at 160 rpm in dim light. Dilute cultures at 4-d intervals with an equal volume of fresh MS-B, and transfer 30 mL of diluted cells to a fresh, autoclaved flask.

2. Transfer BMS cells maintained as above 2 d before electroporation (rather than 4 d). The 1:1 transfer 2 d before electroporation ensures that cells will be in a logarithmic growth phase, which is optimal for high levels of transient gene expression. One flask of cells will, on average, produce enough cells (~10^7) for ~10 electroporations.

3. Transfer the BMS cells to a sterile, 50-mL plastic screw-cap tube. Centrifuge in a clinical-type swinging bucket centrifuge for 4 min at 150g to pellet the cells. Transfer the clear supernatant (i.e., conditioned MS-B medium) to a sterile Nalgene filtration unit (0.45-µm pore size; Nalge, Rochester, NY), and reserve for use in PGM. Aspirate and discard any remaining culture media.

4. To prepare protoplasts, pool the cells from two to three 30-mL culture flasks in ~45 mL of PIM + enzymes. Distribute the cells in Petri plates (~12 mL/plate), and mix gently on a shaker platform at 50 rpm at 25°C. As the digestion progresses, individual, round protoplasts, some with residual cell-wall fragments, will be liberated from larger, irregularly shaped cell clumps. Monitor the progress of the digestion by examination of a plate of cells about every 30 min. Stop the digestion when ~60% or more of the cells exist as round, isolated protoplasts (*see* Note 5).

5. Pool the protoplasts in sterile, 50-mL plastic screw-cap tubes, and centrifuge for 4 min at 150g. Carefully aspirate the enzyme mix from the protoplast pellet.

6. Wash the protoplasts in 30–40 mL of PIM by gently inverting the tube to mix the cells completely, centrifuging as above, and aspirating the liquid. Wash the protoplasts twice with PIM and once with POR. Determine the number of protoplasts per milliliter using a hemocytometer, and pellet cells a final time. Resuspend the pellet in POR for a final protoplast density of 2 × 10^6/mL.

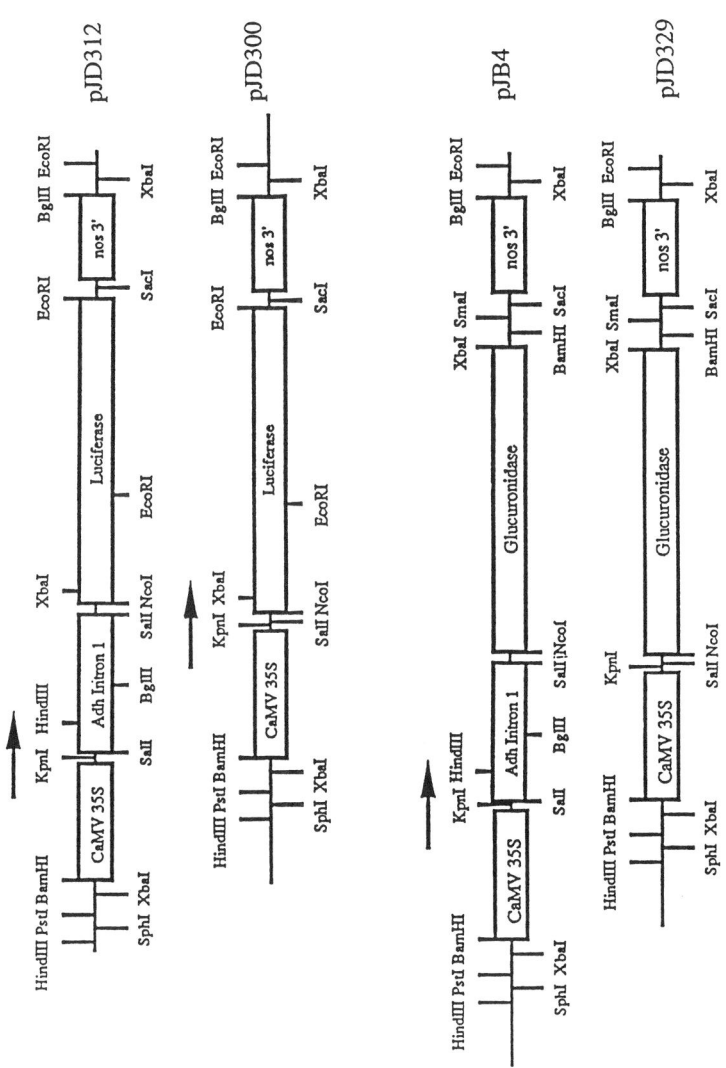

7. For maximal expression of reporter genes, heat-shock the protoplasts for 10 min at 45°C (*see* Note 6). Let the protoplasts recover on ice for 20 min before electroporation.

8. Mix plasmid DNAs for electroporation while the cells digest (*see* Note 7). Include in each electroporation 5–50 μg of an LUC reporter construct (20 μg being average; the amount being constant between all samples), and either 5 or 10 μg of a GUS-expressing plasmid for use as an internal standard. The DNA plasmids are combined and the volume brought up to 500 μL with POR. The 500-μL aliquots are dispensed into 1-mL plastic cuvets (*see* Note 8) and kept in an ice slurry.

9. Before the initial electroporation and between each subsequent one, rinse the electrodes with water, flame-sterilize in ethanol, and cool.

10. Immediately before electroporation, add 500 μL of protoplasts (~10^6) to cuvets containing the 500 μL of plasmid in POR mix made earlier. Keep cuvet on ice during electroporation. Pulse with 200 V at 1550-μF capacitance, with a time constant ≤12 ms. Electroporated protoplast samples darken somewhat in color, and often, bubbles form on top of the cell suspension after electroporation (*see* Note 9).

11. Immediately after electroporation, the cells are gently mixed with 8 mL PGM (with conditioned media added) in a standard Petri dish. Incubate the protoplasts without shaking at 25°C in dim light for 24–48 h under sterile conditions. Although levels of gene expression are detectable within 6–8 h, we generally harvest cells at ~16 h for mRNA isolation, or at ~40 h for LUC and GUS assays for maximal reporter gene expression (*see* Notes 10 and 11). Expression of the reporter genes decreases after this time owing to DNA plasmid degradation, and LUC mRNA and protein turnover.

Fig. 1. *(previous page)* Restriction maps and general structures of LUC and GUS reporter gene expression constructs. The name of each plasmid is shown to the right of each map. Each insert as shown is between 5 and 6 kb. The cauliflower mosaic virus (CaMV) 35S promoter, a high-level constitutive promoter, can be used to drive either LUC or GUS expression; typically, the promoter of the LUC constructs is replaced with the promoter of interest, with the 35S promoter driving levels of the internal GUS plasmid. The Adh intron 1 in the reporter gene construct can enhance reporter gene expression; however, expression levels can be compared with or without an intron. Both LUC and GUS constructs are in the base vector pUC18. Both LUC and GUS vectors are available from either Promega Biotech or Clontech.

3.2. *Transient Gene Expression Analysis*

The following assay for LUC expression in protoplast extracts is adapted from de Wet et al. *(16),* with the substitution of an improved luciferase extraction buffer (CCLR) and LUC assay reagent (LAR) developed by Promega Corp. This provides a simpler and more sensitive assay, as well as more stable reaction kinetics, and allows the assay to be quantitated with a standard lab scintillation counter if desired, eliminating the need for a luminometer if one is not available.

1. Harvest samples for transient assays by collecting cells from each plate into a prelabeled 15-mL screw-cap tube. Dislodge protoplasts adhering to the Petri dish with a plastic cell scraper. Centrifuge protoplasts for 4 min at 150g, and aspirate the supernatant.
2. Resuspend the protoplast pellet in 400 µL LUC extraction buffer stirring for 10 s. Transfer to 1.5-mL tubes, and place on ice.
3. Prepare a cleared cell extract by centrifuging for 10 min at 12,000g at 4°C. If desired, transfer supernatant to a clean tube, and store at 4°C until ready for enzyme assays (*see* Note 12).
4. Add 5–50 µL of cleared cell extract from each electroporation to a luminometer cuvet, and allow to stand at room temperature for 5–15 min. This incubation allows the LUC enzyme in the cell extract to attain its full level of activity. Pipet 200 µL of LAR into the cuvet just prior to measuring activity (*see* Note 13).
5. Measure LUC activity by setting the luminometer to a 10-s integration, and place the sample cuvet in the chamber. LUC activity detected during the counting period is displayed as the number of photons emitted and is recorded as lu/10 s. Typically, for a constitutive promoter, activity will be on the order of 10^4–10^5 lu/10 s. The activity can range from 10^2–10^6 lu/10 s, however, and some highly active extracts may even require dilution before reading.
6. To measure GUS activity, mix in a microfuge tube 350 µL GUS buffer, 50 µL cleared cell extract, and 100 µL MUG solution. Incubate at 37°C. Remove 100 µL of this solution after 5 min as the first time-point, and two later 100-µL aliquots at 25-min intervals (e.g., at 30 and 55 min). Place each 100-µL time-point immediately after gathering into 900 µL of stop buffer. Vortex diluted aliquots, and store at 4°C until all samples have been taken (*see* Note 14).
7. Set baseline reading of fluorimeter with pure stop buffer, and set to 1000 U with 0.1 mM MU standard. Read entire reaction or a dilution for each time point in quartz cuvets, recording the amount of substrate metabolized per min (MUG → MU) as pmol MU/min.

3.3. Data Analysis

It is important to standardize the LUC activities between different electroporations and when comparing LUC constructs. Factors that may cause variations between electroporations or even between samples within a set of electroporations are differences in protoplast viability, protoplast recovery, and purity between different plasmid preparations. The simplest correction, which is an optional one, can be made by standardizing LUC activity with the total protein in each extract and expressing the results as LUC activity (lu/10 s/µg protein). Total protein can be determined easily using the Bradford reagent (Bio-Rad, Richmond, CA), using BSA as a standard. To correct for variations in gene expression, a more important consideration both between and within electroporations, we correct each LUC reading with the amount of GUS activity in that extract as described above, with the final data expressed as LUC activity/ GUS activity, i.e., (lu/10 s)/(pmol MU/min), followed by the standard deviation. Typically, the GUS activity factor does not alter the overall results within an experiment, but does provide a valuable means of comparing results from experiment to experiment. Averaging of duplicate or triplicate electroporations within experiments adds to the reliability of the final data.

4. Notes

1. The devices we have used for our experiments are listed, but equivalent instruments are also available. For protoplast electroporation, we use the Promega X-Cell 450 electroporation device (Promega Biotech); however, this machine is no longer on the market. Two similar electroporators with equivalent electrical parameters are the Hoeffer ProGenetor II and the Bio-Rad Gene Pulser. To measure LUC activity, we use a Monolight 2001 luminometer (Analytical Luminescence); however, using the improved LAR we describe in this chapter, a standard lab scintillation counter will also suffice. To measure GUS activity, we use a Hoeffer TKO-100 DNA minifluorimeter, a low-cost version of a standard laboratory fluorimeter.

2. BMS, available from the American Type Culture Collection (Rockville, MD) (ATCC 54022), forms a fine suspension in liquid culture, grows vigorously, and readily yields protoplasts that reform cell walls and divide. This cell line, however, does not have the capacity to regenerate into maize plants.

3. On autoclaving, MSB and PGM turn a pale yellow color. Over time, a fine, pale yellow or white precipitate may form at the bottom of the bottles con-

taining these solutions. This is most likely owing to oxidized iron (ferric ions). This precipitate, however, does not decrease the effectiveness of the media and need not be filtered out of the solution. We simply shake the bottles to disperse the precipitate before dispensing into BMS culture flasks or protoplast samples.

4. Most commercial preparations of protoplasting enzymes are somewhat crude in purity, and can show variations in the rate of digestion and the subsequent viability of the protoplasts. It is best to compare lots from different vendors and then purchase a sufficient quantity of those lots that performed well. Some investigators treat PIM + enzymes at 55°C for ~10 min before filtering both to inactivate proteases that may be present as well as to aid in solubilization of the enzymes.

5. No attempt should be made to digest cells to completion, since this will lead to loss of protoplast viability. The exact time of digestion varies with initial cell density and freshness of the enzyme mix; a freshly prepared enzyme will often digest cells within 2 h, whereas use of a previously frozen and thawed enzyme mix will require 3–4 h.

6. The 45°C heat-shock of the protoplasts just before electroporation is optional, but has been shown to increase expression levels of the reporter genes several-fold, possibly by inducing a heat-shock response, which confers a degree of protection against the effects of electroporation *(6)*.

7. In the past, we have used plasmids purified by cesium chloride banding as described in Maniatis et al. *(17)*; however, with the advent of faster "column" methods of plasmid preparation (i.e., Qiagen, Chatsworth, CA; Wizard Maxiprep, Promega Biotech), many investigators have switched to using this method of plasmid purification. Although the plasmids obtained from these two methods should be equivalent, it is advisable to compare expression side-by-side from identical plasmids prepared by different methods or to prepare all plasmids by one method only.

8. We use 1-mL disposable spectrophotometric cuvets (cat. no. 58017-847; VWR Scientific, Philadelphia PA) for electroporation, which we load into test tube racks (20–30/rack) and sterilize with ethylene oxide gas, a service available at most hospitals. However, individually wrapped presterilized cuvets can also be used (for example, with the Bio-Rad Gene Pulser), or the assays can be adapted to sterile 24-well dishes, using the ring electrode available on the Hoeffer ProGenetor II.

9. Estimations of protoplast viability can be determined using vital dyes or simply by looking for the presence of cytoplasmic streaming in electroporated cells. Fluorescein diacetate (FDA) or neutral red *(18)* are two commonly used dyes that work well for protoplast viability determination.

Living cells will cleave acetate from FDA, producing fluorescein. On excitation with UV light (excitation λ = 450–490 nm, emission cutoff λ = 520 nm), fluorescein fluorescence will be detected in living cells. A stock solution of FDA in acetone (0.5 mg FDA/mL acetone) is prepared and diluted 1:200 with PGM. Ninety microliters of this solution are added to 10 µL of protoplasts and observed for fluorescence. A simpler procedure, neutral red staining, requires only the use of a light microscope. A 0.01% solution of neutral red in PGM is prepared and mixed 1:1 with a small sample of protoplasts. Living cells will actively concentrate neutral red into the vacuole and will actually swell in size, whereas dead or damaged cells simply remain red throughout.

10. The most important factor determining expression levels of reporter genes and reproducibility between experiments is the physiological state of the BMS cells. The cells must be growing logarithmically for best results. Both the extent of protoplasting and the amount of electrical current during electroporation must be carefully monitored to achieve a balance between effectiveness of gene transfer and protoplast viability. These parameters must be determined empirically for cells other than BMS. For electroporators mentioned above, manufacturer's guidelines are provided. For additional information, the excellent review by Fromm et al. *(4)* covers the many aspects involved in determining electrical parameters for electroporation.

11. The purpose of many transient-expression assays is to examine the expression of one particular construct as a function of different experimental treatments (i.e., increasing temperature, different levels of hormones, and so forth). To do this type of experiment, the same DNA construct (and GUS internal control) is electroporated for as many treatments as is needed, and as above, individually transferred to Petri dishes. On the next morning (12–14 h after electroporation), after cells have recovered from the electroporation and the plasmid DNA has been translocated to the nucleus, the experimental treatments begin (i.e., treatments at different temperatures or hormone concentrations). Depending on the experiment, protoplasts can be recovered for reporter gene assays immediately after treatment or at different time-points. The timing of treatments after electroporation and harvesting of protoplasts for reporter gene assays must be empirically determined for individual experiments (i.e., time-course of initial and final expression).

12. Ideally, the GUS and LUC reporter gene assays should be performed immediately following the preparation of the cleared extract; however, we have found that both LUC and GUS enzyme activity remain relatively stable for approx 1 mo if the cell extract is stored at –80°C.

13. Using the LAR buffer, light emission is constant and stable for approx 5 min. The extended light emission of luciferin using the LAR buffer allows the LUC assays to be quantitated in a scintillation counter. Maximal sensitivity requires turning off the coincidence counter to ensure that all photons are counted as events.
14. GUS-Light, a chemiluminescent GUS assay recently developed by Tropix (Bedford, MA), allows GUS activity to be quantitated easily with a luminometer, allowing both LUC and GUS activity to be measured on one machine.

References

1. DeCleene, M. and DeLey, G. (1976) The host range of crown gall. *Botan. Rev.* **42,** 389–466.
2. Schlappi, M. and Hohn, B. (1992) Competence of immature maize embryos for *Agrobacterium*-mediated gene transfer. *Plant Cell* **4,** 7–16.
3. Fromm, M., Taylor, L. P., and Walbot, V. (1985) Expression of genes transformed into monocot and dicot plant cells by electroporation. *Proc. Natl. Acad. Sci. USA* **82,** 5824–5828.
4. Fromm, M. E., Callis, J., Taylor, L. P., and Walbot, V. (1987) Electroporation of DNA and RNA into plant protoplasts. *Methods Enzymol.* **153,** 351–366.
5. Jefferson, R. A. (1987) Assaying chimeric genes in plants: the GUS gene fusion system. *Plant Mol. Biol. Rep.* **5,** 387–405.
6. Luehrsen, K. R., de Wet, J. R., and Walbot, V. (1992) Transient expression analysis in plants using firefly luciferase reporter gene. *Methods Enzymol.* **216,** 397–414.
7. Marrs, K., Casey, E. S., Capitant, S. A., Bouchard, R. A., Dietrich, P. S., Mettler, I. J., and Sinibaldi, R. M. (1993) Characterization of two maize HSP90 heat shock protein genes: expression during heat shock, embryogenesis, and pollen development. *Dev. Genet.* **14,** 27–41.
8. Pla, M., Vilardell, J., Guiltinan, M. J., Marcotte, W. R., Niogret, M. F., and Quatrano, R. S. (1993) The cis-regulatory element CCACGTGG is involved in ABA and water stress responses of the maize gene rab28. *Plant Mol. Biol.* **21,** 259–266.
9. Bodeau, J. P. and Walbot, V. (1992) Regulated transcription of the maize Bz2 promoter in electroporated protoplasts requires the C1 and R gene products. *Mol. Gen. Genet.* **233,** 379–386.
10. Luehrsen, K. R. and Walbot, V. (1992) Insertion of non-intron sequences into maize introns interferes with splicing. *Nucleic Acids Res.* **20,** 5181–5188.
11. Pitto, L., Gallie, D. R., and Walbot, V. (1992) Role of the leader sequence during thermal repression of translation in maize, tobacco, and carrot protoplasts. *Plant Physiol.* **100,** 1827–1833.
12. Gallie, D. R., Lucas, W. J., and Walbot, V. (1989) Visualizing mRNA expression in plant protoplasts: factors influencing efficient mRNA uptake and translation. *Plant Cell* **1,** 301–311.
13. Gallie, D. R., Feder, J. N., Shimke, R. T., and Walbot, V. (1991) Post-transcriptional regulation in higher eukaryotes: the role of the reporter gene in controlling expression. *Mol. Gen. Genet.* **228,** 258–264.

14. Leon, P., Planckaert, F., and Walbot, V. (1991) Transient gene expression in protoplasts of *Phaseolus vulgaris* isolated from a cell suspension culture. *Plant. Physiol.* **95,** 968–972.
15. Planckaert, F. and Walbot, V. (1989) Transient gene expression after electroporation of protoplasts derived from embryogenic maize callus. *Plant Cell Rep.* **8,** 144–147.
16. de Wet, J. R., Wood, K. V., Helinski, D. R., and DeLuca, M. (1985) Cloning of firefly luciferase and the expression of active luciferase in *Escherichia coli. Proc. Natl. Acad. Sci. USA* **82,** 787–791.
17. Sambrook, J., Fritsch, E. F., and Maniatis, T., eds. (1989) Plasmid vectors, in *Molecular Cloning: A Laboratory Manual.* Cold Spring Harbor Laboratory, Cold Spring Harbor, NY, pp. 1.42–1.50.
18. Gahan, P. B. (ed.) (1984) *Plant Histochemistry and Cytochemistry: An Introduction.* Academic, London.

CHAPTER 11

Reporter Genes
and Transient Assays for Plants

*Benjamin F. Matthews,
James A. Saunders, Joan S. Gebhardt,
Jhy-Jhu Lin, and Susan M. Koehler*

1. Introduction

The introduction of DNA into plant cells, protoplasts, or intact tissues has been accomplished by a variety of mechanisms, including electroporation, electrofusion, particle bombardment, liposome transfer, the use of bacterial vectors, polyethylene glycol treatment, and other procedures. As new techniques are developed or modified, it is necessary to use a reliable gene-expression system to monitor DNA uptake, transcription, and translation. A series of DNA plasmids containing reporter genes encoding readily assayed enzymes are available for this purpose. Several reporter gene systems have been used in experiments to transform plants and to perform transient assays with plant material. In general, these reporter genes encode enzymes whose activities can be detected through assays and stains, thus facilitating the identification of transformed cells and quantification of the transformation process. Reporter genes also provide a method for analyzing regulatory characteristics of promoters, such as promoter strength and tissue specificity, when the promoter from a gene of interest is coupled to the reporter gene.

Reporter gene constructs are comprised of a reporter gene together with active promoter and terminator regions cloned into a plasmid vector. Sometimes these constructs lack promoter regions, so different pro-

From: *Methods in Molecular Biology, Vol. 55: Plant Cell Electroporation and Electrofusion Protocols* Edited by: J. A. Nickoloff Humana Press Inc., Totowa, NJ

moter regions can be inserted at the 5' end of the reporter gene using conveniently located restriction sites. The promoters and pieces of promoters can be inverted, rearranged, modified, and tested to determine how and why the promoter works the way it does. Functions of the promoter regions can be examined by observing expression of the reporter gene in transformed tissue.

Reporter genes are usually incorporated into fairly small plasmid constructs of 4–8 kbp capable of replication in *Escherichia coli* and are easy to handle. It is desirable for the product of the reporter gene to be produced as a stable protein that does not interfere with cell growth or metabolism. An ideal reporter gene would encode an enzyme that could be assayed with commercially available, inexpensive, nontoxic reagents. However, in reality this is not always the case. Detection of enzymes encoded by some genes requires assays that are cumbersome, use radioactively labeled substrates, are time-consuming, or have interfering background reactions. Ideally, the reporter gene product should be detectable in minute quantities with no interfering background. It is sometimes convenient to have a histological stain available for the reporter gene, so temporal and spatial distribution of the product can be examined within plant tissues. The enzyme encoded by the reporter gene must retain activity when aminoterminal fusions are made to study the regulation of gene promoters.

Reporter genes and their enzyme products are used to optimize transformation protocols, and to differentiate between transformed and untransformed cells. In these cases, the reporter gene is controlled by a strong constitutive promoter. Reporter genes can also be used to observe the expression of gene promoters to determine the control patterns of the promoter throughout a developmental sequence, thus furnishing insights into temporal and spatial expression of genes, including tissue specificity. Regulation of gene transcription can be monitored by linking the promoter sequence with the reporter gene, inserting the construct into plant cells or tissue, and examining in vivo gene expression. DNA sequences suspected of having a role in transcriptional regulation, including untranslated leader sequences, can be added, deleted, inverted, modified, or scrambled, and the effects of these modifications can be observed by using reporter enzyme assays. Thus, the effects of these additions, deletions, or other modifications can be quantified by assaying the reporter gene product.

There are several reporter genes that have assumed a prominent role in plant cell transformation studies. These genes encode the reporter enzymes β-glucuronidase (GUS), chloramphenicol acetyltransferase (CAT), luciferase (LUC), and neomycin phosphotransferase (NPTII). There are numerous protocols for using each of these reporter gene systems, and a variety of promoter constructs have been used in conjunction with these genes. The protocols presented here are well characterized. However, they may need to be modified for optimal performance in different plant tissues and in diverse experimental designs. Many manufacturers provide detailed booklets on these assays and provide technical assistance.

The *npt*II gene derived from bacterial transposon Tn5 is a widely used selectable marker incorporated in many plant transformation vectors. The *npt*II gene encodes neomycin phosphotransferase II (NPTII), which inactivates the aminoglycoside antibiotics kanamycin and geneticin (G418) by phosphorylation. Selection for kanamycin resistance allows for easy identification of transformants in many plant species, including tobacco, tomato, rice, wheat, and canola.

Kanamycin resistance may be checked by a callus induction assay and can be followed in the progeny of transgenic plants by sowing seeds on kanamycin-containing medium (NPTII seedling assay). NPTII levels in transgenic tissues can be measured using radiolabeled kanamycin or by enzyme-linked immunosorbent assay (ELISA).

The *uid*A gene *(1)* is one of the most commonly used reporter genes to monitor gene expression and plant transformation. The *uid*A gene encodes GUS and was originally isolated from *E. coli (2)*. Enzymatically active GUS is a tetrameric enzyme composed of 68,200-Dalton subunits. The two most popular GUS assays available are the fluorometric and the histochemical assays, each using different substrates. Of these, the fluorometric assay is more sensitive and is used most often for quantification of GUS expression. A spectrophotometric assay for GUS activity has been used successfully. However, its sensitivity is <.01 that of the fluorometric assay. The spatial distribution of expressed GUS can be monitored using the histochemical assay. The use of GUS is popular, in part, because the substrates for each of these assays are commercially available and do not involve radioactivity. Transit peptides and signal peptides can be used in conjunction with GUS, because it maintains activity when large aminoterminal sequences are added.

Table 1
Comparison of a Chromatographic Assay Using
Radioisotopic Labeled Chemicals and an ELISA Assay for the Detection of CAT

Feature	Thin-layer chromatography assay	ELISA assay
Measurement of CAT	Functional enzyme assay, requires separation of product from substrate	Antibody–antigen interaction
Assay time	24–36 h	3–5 h
Requirement for radioisotopic labeled chemicals	Yes	No
Qualitative assay	Yes, requires autoradiogram for viewing measurement	Yes, color reaction, no autoradiogram
Quantitative assay	Yes, requires expensive equipment, like scintillation counter	Yes, requires inexpensive plate reader
Detection limit	0.5–5.0 pg	20–50 pg

In some plant systems, endogenous enzyme activity may cause low levels of positive activity, particularly during periods of long substrate incubation. However, these endogenous activities are usually associated with contaminating bacteria on the plant tissues, and the GUS reporter system works well in most plant systems.

The gene encoding CAT is derived from bacteria and confers resistance to chloramphenicol. Chloramphenicol is inactivated by acetylation at one or both of its two hydroxyl groups. This gene reporter system is useful because it can be expressed in both mammalian and plant cells. In addition, no detectable CAT activity was found in most plant tissues assayed *(3,4)*.

Traditionally, detection of CAT was performed using a functional enzyme assay with radioisotopically labeled chloramphenicol *(5)*. The assay involves using radioisotopic labeled chemicals, organic solvent extraction, thin-layer chromatography, and autoradiography. It is tedious, time-consuming, and hazardous handling the radioisotopic chemicals and the related chemical wastes. These complicated procedures limit the use of CAT as a reporter gene (Table 1).

Recently, nonradioisotopic CAT assays using either fluorescent substrates *(6)* or an antibody-based ELISA assay have been described *(7)*.

However, using fluorescent substrates for detecting CAT still requires the use of thin-layer chromatography, and requires expensive detection equipment for measuring and analyzing the data. The ELISA is suitable for both qualitative and quantitative detection of CAT activity, and avoids using radioisotopic chemicals, thin-layer chromatography, and autoradiography *(7,8)*. Therefore, the ELISA assay has now become attractive for the detection of CAT expressed in transgenic plants. Here we describe a nonradioisotopic antibody-based ELISA assay for detection of CAT in transgenic plants. In this assay, monoclonal antibody (MAb), which can specifically bind the desired CAT antigen, is bound to the microwell surface. Following incubation in the presence of sample, a primary antibody that can also recognize the CAT antigen is added to the well. Finally, a second antibody that is labeled with horseradish peroxidase (HRP) is added to the well, and the complex is incubated with the HRP substrate. The colored product can be measured spectrophotometrically. The assay is sensitive, quantitative, and reproducible.

The cDNA encoding luciferase (LUC) was cloned from the firefly, *Photinus pyralis (9)*. The LUC assay is dependent on light production and is 1000-fold more sensitive than the standard CAT assay *(10,11)*. Owing to the sensitivity of the LUC assay, fewer cells need to be transformed, and less material needs to be sacrificed to detect positive transformation than with the standard CAT assay *(12)*. The in vitro LUC assay is rapid and easy to perform. It does not require radioisotopes or other hazardous substrates, and except for the cost of a luminometer, it is relatively inexpensive to perform.

The LUC gene has been used as a reporter gene to monitor promoter activity in both prokaryotic and eukaryotic cell lines, including bacteria *(13,14)*, yeast *(15)*, animal cells *(16–18)*, insects *(19,20)*, and plants *(21–23)*. In plants, the LUC reporter gene has some advantages over GUS, because the assay is more sensitive, and unlike GUS, there is no endogenous LUC activity in plants that can produce high backgrounds in nontransformed tissue.

Firefly LUC catalyzes the ATP-dependent oxidative decarboxylation of the synthetic substrate D-luciferin to produce light *(24,25)*. Under standard assay conditions, a flash of light is produced within 0.5 s. It decays rapidly (within 10 s), because the enzyme is bound to the product, oxyluciferin. The light output can be most efficiently detected by a luminometer. The enzyme follows Michaelis-Menten kinetics. Therefore, maximum light intensity will be reached when the substrate concentrations are in excess. In this case, emitted light is directly proportional to the number of LUC enzyme molecules.

2. Materials

2.1. Protoplast Transformation

HEPES-buffered saline (HBS): consists of 10 mM HEPES, pH 7.2, 150 mM KCl, and 4 mM CaCl$_2$, supplemented with sufficient mannitol to stabilize the protoplasts.

2.2. NPTII Assays

Callus induction medium consists of complete Murashige and Skoog (MS) medium (*26*; Sigma Chemical Co., St. Louis, MO) supplemented with 1 mg/L NAA (naphthalene acetic acid) or 2,4-D (2,4-dichlorophenoxyacetic acid), 0.1 mg/l BAP (6-benzylaminopurine), and 0.8% agar containing different antibiotic concentrations (*see* Note 1). Filter-sterilized hormones and antibiotics should be added to the medium after autoclaving and cooling to 55°C.

2.3. GUS Assays

1. Lysis buffer: 50 mM sodium phosphate buffer, pH 7.0, 10 mM EDTA, 0.1% (v/v) Triton X-100, and 10 mM β-mercaptoethanol.
2. GUS histochemical stain: 1–2 mM 5-bromo-4-chloro-3-indolyl glucuronide (X-Gluc) in DMSO, 100 mM potassium phosphate buffer, pH 7.0, 10 mM EDTA, 0.5 mM potassium ferricyanide, 0.5 mM potassium ferrocyanide, and 0.1% Triton X-100.

2.4. CAT Assays

1. CAT reaction mix: 50 μL 1M Tris-HCl, pH 7.8, 10 μL 60 mCi/mmol, ^{14}C-labeled chloramphenicol, 20 μL 3.5 mg/mL acetyl coenzyme A, freshly made.
2. CAT lysis buffer: 50 mM sodium phosphate buffer, pH 7.0, 10 mM EDTA, 0.2% Triton X-100, and 10 mM β-mercaptoethanol.

2.5. LUC Assay

1. LUC cell lysis reagent: 0.1M phosphate buffer, pH 7.8, 1% Triton X-100, 2 mM EDTA, 1 mM DTT.
2. LUC assay buffer solution: 30 mM Tricine, pH 7.8, 3 mM ATP, 15 mM MgSO$_4$, 10 mM DTT.

3. Methods

3.1. Transformation
of Plant Protoplasts by Electroporation

1. Wash freshly isolated protoplasts in HBS solution and resuspend in HBS at 4×10^6 protoplasts/mL.
2. Transfer an equal volume of protoplasts and plasmid DNA (40 µg/mL dissolved in HBS-mannitol) (*see* Note 2) into electroporation chamber.
3. Subject mixture to an 8–10-ms exponential discharge (*see* Note 3).
4. Incubate for 3 min, and then transfer the protoplasts to a 10-fold volume of culture medium.
5. Protoplasts can be maintained for 24 h after electroporation to examine transient gene expression or be cultured for longer-term selection of stable transformants.

3.2. Neomycin
Phosphotransferase (NPTII) Callus Assay

There are two popular methods of assaying for the presence of NPTII. One method is a callus induction assay and the other is a seedling assay (Section 3.3.). The callus induction assay is conducted as follows:

1. Surface-sterilize leaf tissue from transgenic plants by soaking consecutively in the following solutions:
   a. 70% Ethanol, 1–2 min.
   b. 10% Commercial bleach with a drop of detergent, 10–20 min.
   c. Sterile double-distilled water (ddH$_2$O), several rinses of 2–3 min each.
2. Cut the leaf tissue into 0.5-cm^2 segments.
3. Transfer leaf segments to callus induction medium.
4. Check callus growth after 2–3 wk. Leaf tissue from kanamycin resistant plants will produce callus in 2–3 wk. Callus induction from wild-type sensitive plants should be completely inhibited.

3.3. NPTII Sterile Seedling Assay

1. Sterilize seeds as above.
2. Sow seeds on MS medium containing 10 g/L glucose or sucrose, 50–300 mg/L kanamycin, and 0.8% agar.
3. Check for resistant or sensitive plants after 1 mo. Resistant plants will be green, whereas sensitive plants will be bleached or chlorotic (*see* Note 4).

3.4. β-Glucuronidase (GUS) Assay

There are several methods used for monitoring GUS activity in transgenic plants. One method is a fluorometric assay, which uses the substrate 4-methylumbelliferyl-β-D-glucuronide (MUG) with tissue homogenates as the enzyme source to yield the fluorescent product 4-methyl umbelliferone. The 4-methyl umbelliferone has a peak excitation of 365 nm and peak emission of 455 nm, and may be measured using a variety of relatively inexpensive fluorometers (*see* Notes 5 and 6).

1. Wash plant material thoroughly in ddH$_2$O. Homogenize 500 mg of plant tissue in a mortar and pestle in 1.5 mL lysis buffer at 4°C. If motorized homogenizer is used, prevent the sample from overheating. Process on ice or at 4°C.
2. Centrifuge sample at 5000g for 5 min at 4°C. Remove and retain supernatant.
3. Rehomogenize pellet in an additional 1 mL of lysis buffer. Repeat centrifugation, and combine supernatants.
4. Repeat centrifugation with combined supernatants, and retain supernatant.
5. Typically, 300–500 μL of GUS lysis buffer containing 1 mM MUG are added to 300 μL of enzyme extract and the reaction mixture is incubated at 37°C.
6. Monitor the reaction at 0, 1, 2, and 4 h by taking 100-μL aliquots of the reaction mixture. Immediately stop each reaction by adding 1.9 mL of 0.2M Na$_2$CO$_3$.
7. Measure at excitation of 365 nm and emission of 455 nm using a fluorometer. A control sample from nontransformed plant material is concurrently assayed each day to check for slight background activity. A second control sample containing buffer in place of enzyme is used to zero the fluorometer at 0 time.
8. A standard curve for the GUS assay is constructed each day of the assay using 4-methyl-umbelliferone as a standard (*see* Note 7).

3.5. GUS Histochemical Assay

The GUS histochemical assay has been particularly popular because of the ease of operation, reliability of the assay, and vividness of the blue stain from a positive reaction. The protocols were initially described in the 1950s and have been modified several times since *(27–30)*. The reaction is based on a series of reactions that cleave the glucuronide moiety of the substrate, 5-bromo-4-chloro-3-indolyl glucuronide (X-Gluc), and give a blue indigo dye precipitate at the site of β-glucuronidase activity. Typically, the reaction consists of the substrate, the tissue to be stained, the appropriate buffer, and a catalyst.

1. Cut tissue into pieces of approx 1 cm or less. Leaf disks punched with a cork borer work very well, as do hand-cut leaf sections (*see* Note 8).

2. The tissue may be fixed or not fixed; both yield positive results. To fix the material, wash the plant tissues of both experimental and control samples with sterile water, and fix in 2.5% (v/v) formaldehyde in 100 mM sodium phosphate buffer, pH 7.0, for 4–6 min at room temperature. After fixation, wash the tissues again with sterile distilled water.
3. A stock solution of 20 mM X-Gluc in DMSO is diluted to 1–2 mM X-Gluc in 100 mM phosphate buffer, pH 7.0, and the tissue is incubated for 4–6 h at 37°C (*see* Note 9).

3.6. CAT TLC Assay

Two methods for measuring CAT activity in transgenic plants are by thin-layer chromatography and by ELISA detection. The assay using thin-layer chromatography is presented first.

1. Add cold CAT lysis buffer to plant material, and disrupt by grinding with mortar and pestle on ice.
2. Remove debris by centrifugation for 5 min at 16,000g at 4°C.
3. Inactivate endogenous deacylases by incubating 50 µL of extract for 10 min at 60°C. Store remainder of extract at –20°C.
4. Collect particulate matter by 2-min centrifugation at 16,000g at 4°C. Retain supernatant at 4°C.
5. Set up several reactions. Mix each sample with 80 µL of CAT reaction mix, and incubate reactions at 37°C for varying lengths of time (*see* Note 10). Control reaction does not contain enzyme.
6. Stop reaction by addition of 1 mL ethyl acetate. Mix well, and centrifuge at room temperature at 16,000g for 5 min. The organic phase (upper) contains the acetylated forms of chloramphenicol.
7. Transfer 900 µL of the upper phase to a new tube. Avoid interface and lower phase.
8. Remove ethyl acetate by evaporation under vacuum for 30 min.
9. Redissolve pellet, and wash tube sides in a total volume of 20 µL ethyl acetate.
10. Apply entire 20 µL of each reaction 5 µL at a time to the origin of a TLC plate (25 mm silica gel). Dry origin with a hair dryer after each application. Mark origin on the plate with a soft-lead pencil.
11. Place TLC plate in chromatography chamber containing, and equilibrated overnight with, 200 mL chloroform:methanol (95:5). Close chamber, and allow solvent to migrate 3/4 up the plate.
12. Remove plate, and air-dry at room temperature.
13. For alignment of the plate with X-ray film, place adhesive radioactively labeled dots on the TLC plate. Expose X-ray film directly to the TLC plate for 2–3 d.

14. Develop X-ray film, and align with plate. Usually there are three spots on the film. The slowest-migrating spot represents nonacetylated chloramphenicol. The faster-migrating spots represent two forms of chloramphenicol, each acetylated at a different site.

15. CAT activity may be quantitated by scraping the radioactive areas from the TLC plate into a scintillation vial with a spatula and measuring their radioactivity using a scintillation counter.

3.7. ELISA-Based CAT Assay

1. Prechill a mortar and pestle with liquid nitrogen. Place 0.2–0.3 g of transgenic plant tissue into the mortar, keeping the mortar cold by adding liquid nitrogen. Grind the tissue to a fine powder.

2. Keeping the tissue at 4°C, transfer the tissue powder into a chilled 50-mL screw-cap centrifuge tube, add ELISA extraction buffer (1 mL/1 g; $0.1M$ Tris-HCl, pH 7.8), and resuspend tissue thoroughly in the extraction buffer by vortexing.

3. Carefully transfer suspended tissue into a 1.5-mL microcentrifuge tube. Lyse the tissue by freezing the tube in liquid nitrogen or a dry ice-ethanol bath for 5 min, and then thaw in a 37°C water bath. Vortex thawed tissue.

4. Repeat the freeze–thaw process two more times.

5. Centrifuge tube at $16,000g$ for 20 min at 4°C, transfer the supernatant into a new tube, and use 5–10 µL of the supernatant to determine the protein concentration using the Bradford assay *(31)*.

6. Add 100 µL of 2X diluent buffer to each well of a microwell plate. Allow the plate to stand for at least 5 min at room temperature.

7. Add 100 µL of each standard in duplicate to the wells.

8. Add 100 µL of each sample and negative control sample (containing no enzyme extract) in duplicate to the remaining wells on the plate.

9. Cover the plate with a plate sealer, and incubate at 37°C for 60 min.

10. After the incubation period, wash the plate five times with 1X wash buffer. Squeeze the plate frame, and firmly strike the plate on a stack of dry paper towels to remove excess moisture.

11. Add 200 µL of the working primary antibody to each well.

12. Cover the plate with a plate sealer, and incubate at 37°C for 60 min.

13. After the incubation period, wash the plate five times with 1X wash buffer as in step 10.

14. Add 200 µL of the 1X working conjugate to each well.

15. Cover the plate with a plate sealer, and incubate at 37°C for 60 min.

16. After the incubation period, wash the plate five times with 1X wash buffer as in step 10.

17. Add 200 µL of TMB (3',3',5',5'-tetramethylbenzidine) solution to each well, cover the plate, and incubate in the dark at room temperature for 15 min.
18. Add 100 µL of stop solution (2*N* sulfuric acid) to each well.
19. Read the plate using a microtiter plate spectrophotometer set at 450 nm within 30 min of the addition of the stop solution.
20. Plot the OD values obtained for the standards vs the concentration in ng/mL. Calculate the observed concentration of the samples by extrapolating from the graph (*see* Note 11).

3.8. LUC

A general protocol based on a standard in vitro assay *(32)* is given below. Modified assays can be used to obtain more constant light output by removing the inhibitory product, oxyluciferin, from the enzyme. Several compounds can be successfully used for this purpose, the most common being Coenzyme A (CoA) and inorganic pyrophosphate *(33,34)*. Greater sensitivity may be obtained with the modified assays by using a longer incubation time.

1. Typically a small aliquot of cell culture or plant tissue (<100 mg) is extracted by grinding in a 1.5-mL tube using a micropestle in ~3–10 vol of LUC cell lysis reagent on ice (*see* Note 12).
2. Extracts are clarified by centrifugation for 5 min at 16,000*g* at 4°C (*see* Note 13).
3. Bring all reagents to room temperature.
4. Add 20–100 µL of sample extract to assay cuvet.
5. Add 100 µL of LUC assay buffer solution to sample (*see* Note 14).
6. Place assay cuvet in the luminometer (*see* Note 15), and initiate the reaction by injecting 100 µL of 1 m*M* D-luciferin (pH 6.1–6.5).
7. Measure light output at 560 nm for 10–20 s and record the integrated relative light units (RLU; *see* Notes 16–18).

An alternative method that is a modified assay for constant light output is conducted as follows:

1.–2. Same as basic protocol above.
3. Add 20 µL of 1 m*M* CoA or 1 m*M* inorganic pyrophosphate (PPi) solution.
4. Add 100 µL LUC assay buffer solution.
5. Place assay cuvet in the luminometer, and initiate the reaction by injecting 100 µL of 1 m*M* D-luciferin substrate.
6. Measure light output generally for 10–20 s or up to 1 min, and record the integrated RLU.

4. Notes

1. Kanamycin concentrations of 50–500 mg/L and G418 concentrations of 100–200 mg/L are generally used.
2. Linear or supercoiled plasmid DNA may be used.
3. Pulse parameters killing 50% of the protoplasts is a useful starting point for determining optimal pulse settings for transient-expression and stable-transformation experiments.
4. NPTII is not usually monitored by assay, because the assay is difficult and requires radioactive isotope *(35)*. Furthermore, endogenous plant enzymes having broad substrate specificity interfere with the NPTII assay and limit its sensitivity in plant systems. NPTII is also of limited use because there is no histological stain available for detecting NPTII activity in plant tissues.
5. In our hands, the Model 450 Sequoia-Turner fluorometer (Mountain View, CA) has performed well at an affordable price. In laboratory teaching situations, or when a nonquantitative assay of GUS activity is desired, positive or negative determinations of the reaction mixture can be made by placing 1.5-mL tubes containing the reaction mix on the surface of a UV light, such as a DNA transilluminator. Positive determinations of GUS expression are apparent by a bright blue fluorescence of the reaction mixture in the tubes when compared to control samples. In this type of reaction, the same tubes can be nondestructively monitored numerous times over a several-hour time-course and returned to the incubation chamber. Mention of trademark, proprietary product, or vendor does not constitute a guarantee of warranty of the product by the US Department of Agriculture, and does not imply its approval to the exclusion of other products or vendors that may be suitable.
6. GUS activity is optimum at pH 5.2–8.0, and is thermostable with a half-life of 2 h at 55°C. The GUS assay is often run at 37°C for as long as overnight with stable, linear enzyme activity. Also, GUS maintains enzymatic activity in the presence of many ions and detergents, including EDTA, Triton X-100, and sarkosyl.
7. The GUS assay should be linear from 0–100 ng of 4-methyl-umbelliferone in lysis buffer. Protein determinations based on the Bradford protein assay *(31)* (Pierce Coomassie Protein Assay Reagent Kit, Pierce, Rockford, IL) are used to quantitate the specific activity of the enzyme. Sarkosyl can give positive readings with this protein assay, so the protein standard curve must be adjusted for the contributions of sarkosyl in the lysis buffer or the sarkosyl may be eliminated from the lysis buffer resulting in less-solubilized enzyme extracted.

8. Callus tissue can easily be broken into conveniently sized pieces. Vascular tissues, particularly the veins and midribs of leaves, invariably stain more rapidly than interveinal tissue even when constitutive promoters, such as the CaMV 35S promoter, are used.

9. Overnight incubations in X-Gluc are common, but sometimes result in light-blue background staining of control samples. Raising the pH of the buffer to between 7.5 and 8.0 as well as including 20% (v/v) methanol in the incubation medium helps to eliminate any endogenous background staining *(36,37)*. If chlorophyll interferes with tissue level localization of the GUS stain, the specimens can be bleached with 95% (v/v) ethanol to remove excess chlorophyll.

10. Incubation time is dependent on expressed CAT activity, which varies according to promoter strength, tissue specificity, and other parameters. Time intervals of 1 h through overnight may be used. For overnight incubations, the addition of 10 μL more of acetyl CoA to each reaction after 2 h may increase sensitivity.

11. The ELISA-based CAT assay described here is designed around the use of a commercially available kit from Life Technologies, Inc. (Gaithersburg, MD). Similar kits can be purchased from 5 Prime-3 Prime (Boulder, CO) and Boehringer Mannheim (Indianapolis, IN). Some procedures may need to be modified from different manufacturers. However, the basic principles for signal amplification and detection are similar among these manufacturers.

12. The LUC enzyme is more efficiently extracted, solubilized, and stable in the presence of the nonionic detergent Triton X-100 than in the presence of cationic and zwitterionic detergents. However, lower Triton X-100 concentrations should be used if the extracts are also to be assayed for GUS activity using the 4-MUG fluorometric assay. LUC is more stable in buffers, such as Tris-phosphate, tricine, HEPES, or phosphate, at pH 7.5–8.0. Glycylglycine, which has previously been cited for use in the assay buffer *(38)*, is not recommended. DTT is added to improve enzyme stability, and is also necessary in the modified assay to keep CoA in the reduced form. EDTA is included to chelate heavy metal ions, which can interfere with activity.

13. Transgenic plant material should ideally be extracted fresh and assayed immediately, since the enzyme is extremely sensitive to repeated freezing and thawing. If this is not possible, the tissue or extracts can be frozen in liquid nitrogen and stored at −70°C.

14. Maximum light output is obtained within a relatively narrow range of ATP concentrations above the K_m value, and under most conditions, final concentrations of 0.5–1.0 mM ATP will be optimal. However, since the sample also contributes some ATP to the assay, individual conditions may vary.

Be sure the pH is adjusted appropriately. It is important to maintain the optimum reaction pH of 7.8, because under more acidic conditions, the light output shifts to 616 nm (red light), which is not as efficiently detected by most luminometers.

15. Luminometers with automatic injectors, such as the LKB model 1250 (Bromma, Sweden) and Analytical Luminescence Laboratory Monolight 2010 (San Diego, CA), are sensitive luminometers that are designed appropriately for this assay. The Monolight 2010 is capable of detecting 10^{-19} mol of LUC *(11)*.

16. RLU are usually expressed based on mg fresh weight or mg protein of the sample extract basis.

17. Dehydroluciferin is a competitive inhibitor of luciferin that can form on exposure of D-luciferin to UV light and/or moisture and under alkaline storage conditions. Therefore, it is important to observe the manufacturer's recommended procedures for storage in the dark at −20°C and reconstitution of substrate to a pH of 6.0–6.3. Once reconstituted, the substrate can be stored at 4°C for 2 wk or at −20°C or less for several months.

18. Test several volumes of sample to be sure the assay is linear because inhibitory substances in the sample can decrease activity when large sample volumes are used. The time required for the light output to become constant after addition of luciferin may be delayed by a few seconds in the presence of CoA or PPi. Therefore, it is wise to perform a preliminary assay to monitor the kinetics and then adjust the postinjection delay time accordingly prior to measurement.

References

1. Berlyn, M. (1992) In *E. coli,* it's still "uidA"—not "gusA." *Plant Mol. Biol. Reporter* **10,** 11.

2. Jefferson, R. A., Burgess, S. M., and Hirsh, D. (1986) β-glucuronidase from *Escherichia coli* as a gene-fusion marker. *Proc. Natl. Acad. Sci. USA* **83,** 8447–8451.

3. Herrera-Estrella, L., Depicker, A., Van Montagu, M., and Schell, J. (1983) Expression of chimeric genes transferred into plant cells using a Ti-plasmid-derived vector. *Nature* **303,** 209–213.

4. DeBlock, M., Herrera-Estrella, L., Van Montagu, M., Schell, J., and Zambryski, P. (1984) Expression of foreign genes in regenerated plants and in their progeny. *EMBO J.* **3,** 1681–1689.

5. Gorman, C. M., Moffat, L. M., and Howard, B. H. (1982) Recombinant genomes which express chloramphenicol acetyltransferase in mammalian cells. *Mol. Cell Biol.* **2,** 1044–1051.

6. Young, S. L., Barbera, L., Kaynard, A. H., Haugland, R. P., Kang, H. C., Brinkley, M., and Melner, M. H. (1991) A nonradioactive assay for transfected chloramphen-

icol acetyltransferase activity using fluorescent substrates. *Anal. Biochem.* **197,** 401–407.

7. Burns, D. K. and Crowl, R. M. (1987) An immunological assay for chloramphenicol acetyltransferase. *Anal. Biochem.* **162,** 399–404.

8. Gendloff, E. H., Bowen, B., and Buchholz, W. G. (1990) Quantitation of chloramphenicol acetyltransferase in transgenic tobacco plants by ELISA and correlation with gene copy number. *Plant Mol. Biol.* **14,** 575–583.

9. deWet, J. R., Wood, K. V., Helinski, D. R., and DeLuca, M. (1985) Cloning of firefly luciferase cDNA and the expression of active luciferase in *Escherichia coli. Proc. Natl. Acad. Sci. USA* **82,** 7870–7873.

10. Gould, S. J. and Subramani, S. (1988) Firefly luciferase as a tool in molecular and cell biology. *Anal. Biochem.* **175,** 5–13.

11. Subramani, S. and DeLuca, M. (1988) Applications of the firefly luciferase as a reporter gene. *Gen. Eng.* **10,** 75–89.

12. Fulton, R. and Van Ness, B. (1993) Luminescent reporter gene assays for luciferase and β-galactosidase using a liquid scintillation counter. *BioTechniques* **14,** 762,763.

13. Palomares, A. J., DeLuca, M., and Helinski, D. (1989) Firefly luciferase as a reporter enzyme for measuring gene expression in vegetative and symbiotic *Rhizobium meliloti* and other gram-negative bacteria. *Gene* **81,** 55–64.

14. Wolk, C. P., Cai, Y. P., and Panoff, J. M. (1991) Use of a transposon with luciferase as a reporter to identify environmentally responsive genes in a cyanobacterium. *Proc. Natl. Acad. Sci. USA* **88,** 4438–4442.

15. Aflalo, C. (1990) Targeting of cloned firefly luciferase to yeast mitochondria. *Biochemistry* **29,** 4758–4766.

16. Brasier, A. R., Tate, J. E., and Habener, J. F. (1989) Optimized use of the firefly luciferase assay as a reporter gene in mammalian cell lines. *BioTechnology* **7,** 1116–1122.

17. deWet, J. R., Wood, K. V., DeLuca, M., Helinski, D. R., and Subramani, S. (1987) Firefly luciferase gene: structure and expression in mammalian cells. *Mol. Cell. Biol.* **7,** 725–737.

18. Nguyen, V. T., Morange, M., and Bensaude, O. (1988) Firefly luciferase luminescence assays using scintillation counters for quantitation in transfected mammalian cells. *Anal. Biochem.* **171,** 404–408.

19. Hasnain, S. E. and Nakhai, B. (1990) Expression of the gene encoding firefly luciferase in insect cells using a baculovirus vector. *Gene* **91,** 135–138.

20. Jha, P. K., Nakhai, B., Sridhar, P., Talwar, G. P., and Hasnain, S. E. (1990) Firefly luciferase, synthesized to very high levels in caterpillars infected with a recombinant baculovirus, can also be used as an efficient reporter enzyme in vivo. *FEBS Lett.* **274,** 23–26.

21. Ow, D., Wood, K. V., DeLuca, M., deWet, J. R., Helinski, D. R., and Howell, S. H. (1986) Transient and stable expression of the firefly luciferase gene in plant cells and transgenic plants. *Science* **234,** 856–859.

22. Koncz, C., Langridge, W. R., Olsson, O., and Schell, J. (1990) Bacterial and firefly luciferase genes in transgenic plants: advantages and disadvantages of a reporter gene. *Dev. Genet.* **11,** 224–232.

23. Quandt, J. J., Broer, I., and Puhler, A. (1992) Tissue-specific activity and light-dependent regulation of a soybean RBCS promoter in transgenic tobacco plants monitored with the firefly luciferase gene. *Plant Sci.* **82,** 59–70.
24. Deluca, M. and McElroy, W. D. (1978) Purification and properties of firefly luciferase. *Methods Enzymol.* **57,** 3–25.
25. Wannlund, J., DeLuca, M., Stempl, K., and Boyer, P. D. (1978) Use of ^{14}C-carboxyl-luciferin in determining the mechanism of the firefly luciferase catalyzed reactions. *Biochem. Biophys. Res. Comm.* **81,** 987–992.
26. Murashige, T. and Skoog, F. (1962) A revised medium for rapid growth and bioassays with tobacco tissue cultures. *Physiol. Plantarum* **15,** 473–497.
27. Fishman, W. H. (1955) β-Glucosidase. *Adv. Enzymol.* **16,** 361–409.
28. Pearson, B., Andrews, M., and Grose, F. (1961) Histochemical demonstration of mammalian glucosidase by means of 3-(5-bromoidoyl)-β-D-glucopyranoside. *Proc. Soc. Exp. Biol.* **108,** 619–623.
29. Jefferson, R. A. (1987) Assaying chimeric genes in plants: the GUS gene fusion system. *Plant Mol. Biol. Rep.* **5,** 387–405.
30. Jefferson, R. A., Kavanagh, T. A., and Bevan, M. W. (1987) GUS fusions: β-glucuronidase as a sensitive and versatile gene fusion marker in higher plants. *EMBO J.* **6,** 3901–3907.
31. Bradford, M. M. (1976) A rapid and sensitive method for the quantitation of microgram quantities of protein utilizing the principle of protein-dye binding. *Anal. Biochem.* **72,** 248–254.
32. *Luciferase Assay Guide Book, Analytical Luminescence* (1992) 11760 Sorrento Valley Road, Suite E, San Diego, CA 92121.
33. Airth, R. L., Rhodes, W. C., and McElroy, W. D. (1958) The function of coenzyme A in luminescence. *Biochem. Biophys. Acta.* **27,** 519–532.
34. McElroy, W. D. and Seliger, H. M. (1961) Mechanism of bioluminescent reactions, in *Light and Life* (McElroy, W. D. and Glass, B., eds.), Johns Hopkins Press, Baltimore, pp. 219–253.
35. Reiss, B., Sprengel, R., Will, H., and Schaller, H. (1984) A new sensitive method for qualitative and quantitative assay of neomycin phosphotransferase in crude cell extracts. *Gene* **3,** 217–223.
36. Kosugi, S., Ohashi, Y., Nakajima, K., and Arai, Y. (1990) An improved assay for β-glucuronidase in transformed cells: methanol almost completely suppresses a putative endogenous β-glucuronidase activity. *Plant Sci.* **70,** 133–140.
37. Martin, T., Wohner, R. V., Hummel, S., Willmitzer, L., and Frommer, W. B. (1992) The GUS reporter system as a tool to study plant gene expression, in *Gus Protocols, Using the GUS Gene as a Reporter of Gene Expression* (Gallagher, S. R., ed.), Academic, San Diego, pp. 23–46.
38. Howell, S. H., Ow, D. W., and Schneider, M. (1989) Use of the firefly luciferase gene as a reporter of gene expression in plants, in *Plant Molecular Biology Manual* (Gelvin, S. and Schilperoort, R., eds.), Kluwer Academic, Dordrecht, Netherlands, pp. 1–11.

PART III

ELECTROFUSION PROTOCOLS

CHAPTER 12

Electrofusion of Plant Protoplasts

Selection and Screening for Somatic Hybrids of Nicotiana

Harold N. Trick and George W. Bates

1. Introduction

Discoveries in the 1960s and 1970s showed that plant protoplasts could be freed from their walls by digestion with fungal enzymes *(1)*, grown in culture, and regenerated back into intact plants *(2)*. This work opened the way for protoplast fusion and somatic hybridization in plants *(3)*. In the ensuing 20 years, there have been substantial improvements in protoplast fusion and culture, and sophisticated approaches for the selection and characterization of somatic hybrids have been developed. As a result, in the 1990s, protoplast fusion has come of age, and fusion-derived somatic hybrids are being evaluated in practical plant breeding programs *(4–6)*.

Protoplast fusion can be achieved by treatment with polyethylene glycol (PEG) in combination with Ca^{2+} and high pH *(7)* or by electrofusion. Although PEG-induced fusion is often quite effective, it can also be highly cytotoxic. Electrofusion was developed in the early 1980s *(8–10)*, and has been widely applied to the fusion of plant protoplasts *(11)*. Electrofusion has proven to be a rapid, simple, and reproducible technique for the fusion of plant protoplasts from a wide range of plant species and cell types. Two to 10% of the treated protoplasts are fused in a typical electrofusion experiment. In this chapter, we describe the electrofusion, hybrid culture, and selection procedures for the production of interspecific hybrids of *Nicotiana*.

From: *Methods in Molecular Biology, Vol. 55: Plant Cell Electroporation and Electrofusion Protocols* Edited by: J. A. Nickoloff Humana Press Inc., Totowa, NJ

The theoretical basis for electrofusion has been discussed in numerous reviews *(12–14)*. The following overview describes how the fusion is accomplished: The protoplasts are suspended in a solution of low conductivity (usually just mannitol or mannitol plus a small amount of Ca^{2+}) and are introduced into a fusion chamber between parallel wires or plates, which serve as electrodes. A low-voltage (100–200 V/cm), rapidly oscillating AC field (0.5–1.5 MHz) is applied to the electrodes. This field transiently polarizes the plasma membrane surface, causing the protoplasts to line up in chains, which radiate outward from the electrode surface. This cell alignment, which is termed dielectrophoresis, creates the cell-to-cell contacts that are a prerequisite for fusion. Once cell alignment is complete, a short (10–50 μs), high-voltage (1000–2000 V/cm) DC pulse is applied. This pulse electroporates the plasma membranes and causes fusion between protoplasts in contact. After fusion, the cells are simply washed out of the chamber and diluted into culture medium.

Because all cell-fusion techniques leave many cells unfused, successful somatic hybridization depends on the recovery of interspecific hybrids from the population of fused and unfused cells. For species in which low-cell-density culture techniques have been developed, it is possible, by electrofusion, to fuse and culture individual pairs of cells *(15),* or following mass cell electrofusion, cell hybrids can be identified visually, isolated using micropipets, and cultured in microdroplets or on feeder layers. However, protoplasts of most species can only be cultured at high cell densities. In general, therefore, somatic hybridization is critically dependent on the availability of a genetic selection or screening for the identification of protoplast fusion-derived cell hybrids.

A wide variety of genetic markers and mutants have been used for selection and identification of somatic hybrids. The markers chosen will likely depend on the species to be hybridized and the goal of the hybridization experiment. For example, interspecific plastid transfer by somatic hybridization is often achieved by selection for a plastid-encoded deficiency in chlorophyll production, or for plastid-encoded resistance to antibiotics (e.g., streptomycin) or herbicides *(16)*. Also possible is the interspecific transfer of nuclear genes by protoplast fusion *(17,18)*. Particularly useful are selections based on kanamycin resistance (introduced into one fusion partner by *Agrobacterium*-mediated transformation) or a mutation resulting in a nuclear-encoded chlorophyll deficiency. These marker genes can be used in combination with iodoacetate treatment or γ irradiation. For example, when protoplasts of a kanamycin resistant species are mitotically inactivated by treatment with iodoacetate or γ irradiation and fused with untreated protoplasts of a kanamycin-sensitive species, only somatic hybrids will grow on media containing kanamycin (Fig. 1A). Alternatively, iodoacetate-treated or γ-irradiated protoplasts can be fused with protoplasts bearing a nuclear-encoded chlorophyll deficiency, and hybrids can be identified by their ability to turn green on appropriate culture media (Fig. 1B). Iodoacetate inactivates protoplasts by acetylating proteins. This treatment leaves the nuclear genome intact, although organellar genes from the treated cell are often eliminated in the fusion-derived hybrids *(16)*. Low doses of γ- or X-rays can be used to inactivate protoplasts mitotically without substantially damaging the nuclear genome of the treated cell, whereas high doses result in partial chromosome elimination from the treated cell in the hybrids *(18)*. Protocols for iodoacetate and γ-ray treatment of protoplasts, electrofusion, selection of fusion products on kanamycin, and media for greening are described in this chapter.

2. Materials
2.1. Plant Sources

Protoplasts can be isolated from a variety of sources, including hypocotyls, young roots, suspension cultures, and young leaves. For many dicot species, leaf mesophyll protoplasts are readily cultured. However, mesophyll protoplasts from some species, particularly monocots, cannot be cultured. In this case, suspension cultures are usually used for protoplast isolation. We prefer to use leaves for the isolation of tobacco protoplasts because large numbers of leaf mesophyll protoplasts can be harvested and these protoplasts are genetically euploid.

N. tabacum, homozygous for the codominant nuclear-encoded sulfur mutation (genotype Su/Su), can be used as a nongreen fusion partner in protocols where hybrids are identified by greening. Seeds of this genotype are available from the USDA-ARS Crops Research Laboratory, Oxford, NC. When homozygous, the Su mutation results in sulfur-colored plants and calli. This mutation can be complemented by the wild-type allele present in other *Nicotiana* species *(18,19)*. Complementation can be observed in vitro in hybrid calli.

N. plumbaginifolia plants, transformed by *Agrobacterium*-mediated transformation using the binary vector pBI121 (Clonetech; *see* ref. *20)*

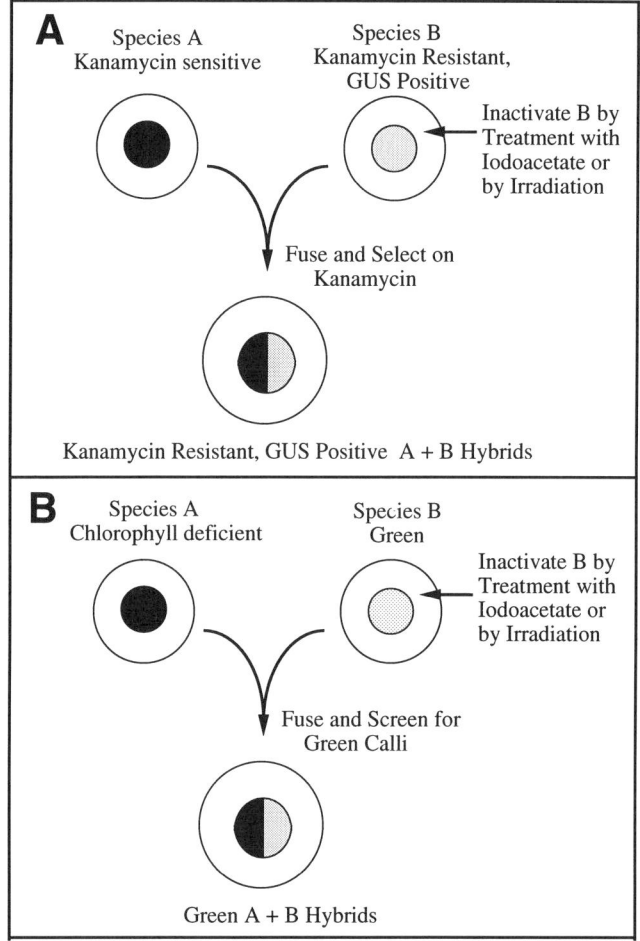

Fig. 1. Outline of selection and screening protocols for recovery of somatic hybrids. In both protocols, protoplasts of the species bearing a dominant-marker gene (i.e., kanamycin resistance or the ability to green in culture) are inactivated by irradiation or treatment with iodoacetate. After inactivation and fusion, only hybrid calli will grow in the presence of kanamycin **(A),** or only hybrid calli will turn green **(B)**.

and the leaf-disk transformation procedure *(21),* are a good source of kanamycin resistant protoplasts. These plants express the neomycin phosphotransferase II gene, encoding resistance to the antibiotic kanamycin, and the reporter gene, β-glucuronidase (GUS). Both of these genes are useful for hybrid selection and identification. *N. plumbaginifolia* also carries the wild-type allele for the Su locus.

2.2. Solutions

1. CPW salts + 0.48M mannitol *(22)*: 1 mM KNO$_3$, 1 mM CaCl$_2$, 1 mM MgSO$_4$, 0.2 mM KH$_2$PO$_4$, 1 mM KI, 604.5 mg/L 2-(N-morpholino) ethanesulfonic acid (MES), 0.48M mannitol. Adjust to pH 5.7 with 1M NaOH. Sterilize by autoclaving. May be kept at room temperature or 4°C.

2. Enzyme solution: 0.15% Cellulysin (Calbiochem, San Diego, CA) and 0.03% pectolyase (Kanematsu Inc., Los Angeles, CA) dissolved in CPW salts + 0.48M mannitol. Adjust to pH 5.2 with 1M NaOH. Filter-sterilize through a 0.22-μm membrane, and use immediately.

3. Fusion medium: 0.5M mannitol, 0.5 mM CaCl$_2$. Filter-sterilize through a 0.22-μm membrane, and store at 4°C.

4. K$_3$ medium + 0.46M glucose *(23)*: 150 mg/L NaH$_2$PO$_4$ · 2H$_2$O, 900 mg/L CaCl$_2$ · 2H$_2$O, 2.5 g/L KNO$_3$, 250 mg/L MgSO$_4$ · 7H$_2$O, 134 mg/L (NH$_4$)$_2$SO$_4$, 0.75 mg/L KI, 40 mg/L Fe/Na EDTA, 10 mg/L MnSO$_4$ · H$_2$O, 2 mg/L ZnSO$_4$ · 7H$_2$O, 3 mg/L H$_3$BO$_3$, 0.25 mg/L Na$_2$MoO$_4$ · 2H$_2$O, 0.025 mg/L CuSO$_4$ · 7H$_2$O, 0.025 mg/L CoCl$_2$ · 6H$_2$O, 10 mg/L thiamine HCl, 1 mg/L nicotinic acid, 1 mg/L pyridoxine HCl, 100 mg/L myo-inositol, 250 mg/L D-xylose, 0.1 mg/L 2,4-D, 0.2 mg/L 6-benzylaminopurine (BA), 1 mg/L α-naphthaleneacetic acid (NAA), and 0.46M glucose. Adjust to pH 5.7 with 1M NaOH. Autoclave and store at 4°C (*see* Notes 1 and 2).

5. TM-2 *(24)*: 18 mg/L Fe/Na-EDTA, 150 mg/L casein hydrolysate, 40 mg/L adenine hemisulfate, 100 mg/L L-glutamine, 68.4 g/L sucrose, 97.6 mg/L MES, 4.56 g/L mannitol, 3.8 g/L xylitol, 4.56 g/L sorbitol, 4.6 g/L myo-inositol, 0.5 mg/L zeatin riboside, 1 mg/L NAA, 170 mg/L KH$_2$PO$_4$, 440 mg/L CaCl$_2$ · 2H$_2$O, 1.5 g/L KNO$_3$, 370 mg/L MgSO$_4$ · 7H$_2$O, 0.25 mg/L nicotinic acid, 1 mg/L thiamine HCl, 0.1 mg/L pyridoxine HCl, 0.05 mg/L folic acid, 0.005 mg/L biotin, 0.05 mg/L D-pantothenic acid, 0.01 mg/L choline chloride, 0.05 mg/L glycine, 0.1 mg/L L-cysteine, 1 mg/L malic acid, 0.05 mg/L ascorbic acid, 0.025 mg/L riboflavin. Adjust to pH 5.7 with 1M NaOH. Filter-sterilize and store at 4°C (*see* Notes 1 and 2).

6. TM-2A: Same as TM-2 but without mannitol, xylitol, sorbitol, and myo-inositol and half of the hormone concentrations (0.25 mg/L zeatin riboside, 0.5 mg/L NAA) (*see* Notes 1 and 2).

7. Murashige and Skoog medium (MS; *see* ref. *25*): 170 mg/L KH$_2$PO$_4$, 440 mg/L CaCl$_2$ · 2H$_2$O, 1.65 g/L NH$_4$NO$_3$, 1.9 g/L KNO$_3$, 370 mg/L MgSO$_4$ · 7H$_2$O, 40 mg/L Fe/Na-EDTA, 22.3 mg/L MnSO$_4$ · H$_2$O, 8.6 mg/L ZnSO$_4$ · 7H$_2$O, 6.2 mg/L H$_3$BO$_3$, 0.25 mg/L Na$_2$MoO$_4$ · 2H$_2$O, 0.025 mg/L CuSO$_4$ · 7H$_2$O, 0.025 mg/L CoCl$_2$ · 6H$_2$O, 0.1 mg/L thiamine HCl, 0.5 mg/L nico-

tinic acid, 0.5 mg/L pyridoxine HCl, 2 mg/L glycine, 100 mg/L myo-inositol, and 3% sucrose. Adjust to pH 5.7 with $1M$ NaOH. Four grams per liter of Phytogel (Sigma Chemical Co., St. Louis, MO), a gelling agent, are added if solid media is needed. Autoclave and store at 4°C (*see* Note 1).

8. CM: Same as MS, but also include 100 mg/L myo-inositol, 1 mg/L BA, and 1 mg/L NAA. Adjust to pH 5.7 with $1M$ NaOH. Variations of this medium, containing different amounts of mannitol, are used in some solutions. Autoclave and store at 4°C (*see* Notes 1 and 2).

9. SeaPlaque agarose medium *(26)*: Dissolve 2.4% SeaPlaque agarose (FMC Corporation, Rockland, ME) in CM + $0.23M$ mannitol, and sterilize by autoclaving. Store at 4°C.

10. Kanamycin stock: 100 mg/mL in water. Filter-sterilize and store at –20°C. Add 1 mL of stock/L of culture medium just before use. Do not add kanamycin when the culture media is above 50°C. Kanamycin in solution is reported to degrade significantly after 1–2 wk at room temperature.

11. MUG solution *(20)*: 100 μM 4-methyl umbelliferyl glucuronide (MUG) (Sigma) in 50 mM sodium phosphate, 10 mM β-mercaptoethanol, 10 mM EDTA, 0.1% sodium lauryl sarcosine, 0.1% Triton X-100, pH 7.0. Store at 4°C. Stable for about 3 mo.

2.3. Electrofusion Instrument and Electrofusion Chamber

Electrofusion instruments are available commercially (suppliers are listed in Chapter 2 and ref. *27*) or can be constructed in-house (*see* refs. *28* and *29* for descriptions). At a minimum, the instrument must be able to produce AC fields of 0–20 V, 0.5–1 MHz, and DC, square-wave pulses of 0–100 V, 10–50 μs. It is also important that the instrument have a circuit that shuts off the AC field during delivery of the DC pulse.

The design of the electrofusion chamber is crucial to success. For effective fusion and hybrid selection, each sample to be fused should contain 2–5×10^5 protoplasts. The fusion chamber should contain 0.25–0.5 mL of protoplast suspension, and the electrodes in the chamber should be 0.4–0.5 mm apart (*see* Note 3). A suitable electrofusion chamber is shown in Fig. 2 *(30)*. This chamber consists of a series of Au/Pd wires (0.2-mm diameter) laid parallel to each other, 0.5 mm apart, on a polycarbonate plastic slide (*see* Note 4). A polycarbonate ring is laid over the wires to form a well that will contain up to 0.5 mL of the protoplast suspension. The wires are soldered to posts mounted on opposite sides of the well. Each wire is soldered to only one post, and adjacent wires are soldered to opposite posts. Thus, each wire is of opposite polarity with

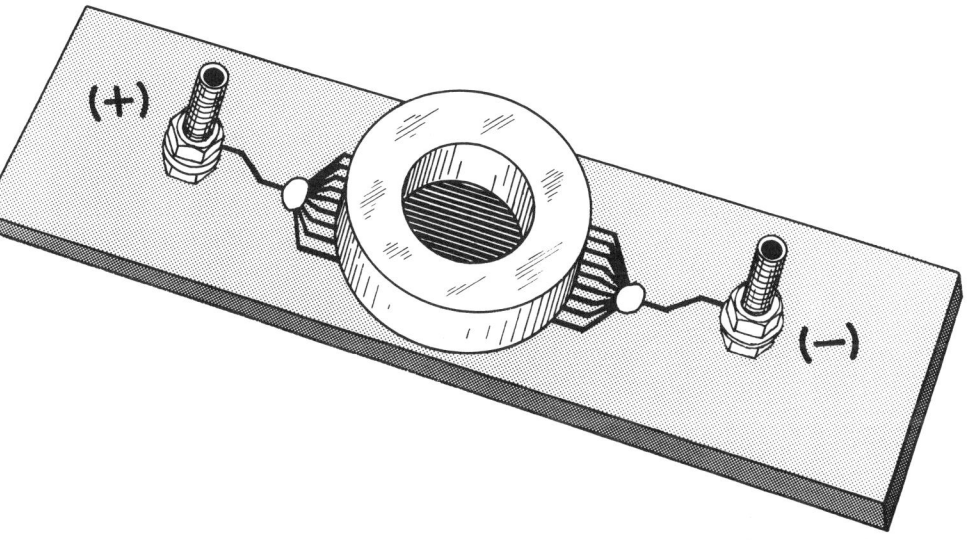

Fig. 2. Drawing of a fusion chamber used for fusion of large volumes (0.25–0.5 mL) of protoplast suspensions *(30)*.

respect to its immediate neighbors. The posts serve as points for connection of leads from the pulse generator to the fusion chamber.

2.4. Radiation Source

A Cs^{137} source with a dose rate of 37 Gy/min is effective. Alternative sources of ionizing radiation include ^{60}Co γ rays or X-rays.

3. Methods

3.1. Maintenance of Source Plants

1. Derive all plant material from either seed or, as with transformed plants, use plants regenerated from leaf disk transformations *(21)*.
2. Propagate plants aseptically by shoot tip transfer on solidified half-strength MS media + 1.5% sucrose in Magenta vessels (Sigma) at 27°C, 20–60 $\mu E/m^2$ (16 h light; 8 h dark).
3. Transfer plants every 4 wk.

3.2. Mesophyll Protoplast Isolation

The protocol below is used to isolate protoplasts from tobacco leaves from both *N. tabacum* and *N. plumbaginifolia*. All procedures are carried out in a laminar flow hood using aseptic techniques.

1. Using a scalpel, cut sterile leaves into thin strips (2–5 mm wide), and float the strips in separate 100 × 20 mm Petri plates on the enzyme solution.

Digest leaf strips for 12–16 h (overnight) at room temperature with 15 min of gentle shaking after 8 h of digestion (*see* Note 5).

2. After digestion, filter the protoplasts through a 70-μm mesh nylon screen (Small Parts Inc., Miami, FL), pipet into 50-mL sterile, conical centrifuge tubes, and pellet at 50*g* for 5 min in a swinging bucket centrifuge. Gently resuspend protoplasts in 10 mL of CPW salts + 0.48*M* mannitol, transfer to 15 mL sterile, conical centrifuge tubes, and pellet at 50*g*. Wash protoplasts once more in 10 mL of CPW salts + 0.48*M* mannitol (*see* Note 6).

3. Resuspend the pellet in 6 mL of filter-sterilized, 18% sucrose. Overlay the resuspended protoplasts with 2–3 mL of CPW salts + 0.48*M* mannitol, allow protoplasts to stand for 10 min, and centrifuge at 100*g* for 5 min (*see* Note 7).

4. Harvest the band of protoplasts at the top of the sucrose gradient with a Pasteur pipet, taking as little of the sucrose pad as possible, and dilute to 10 mL with CPW salts + 0.48*M* mannitol. Pellet protoplasts at 50*g* for 5 min, resuspend in 10 mL of CPW salts + 0.48*M* mannitol, and determine the yield of protoplasts using a hemocytometer. These protoplasts are now ready for fusion (Section 3.5.).

3.3. γ Irradiation

γ irradiation is one method for mitotic inactivation of protoplasts. For *N. plumbaginifolia,* a dose of 50 Gy (*see* Note 8) is sufficient to completely prevent mesophyll protoplasts from forming colonies (5–10% of the protoplasts may divide once, but multiple divisions and colony formation are blocked). Radiation treatment is given prior to the sucrose flotation (Section 3.2., step 3).

1. Pellet protoplasts at 50*g*, and resuspend in 5–8 mL of CPW salts + 0.48*M* mannitol and transfer to a 60 × 15-mm Petri plate.
2. Expose protoplasts to 50 Gy from a γ- or X-ray source.
3. Transfer the protoplasts to a 15-mL centrifuge tubes and pellet (50*g*).
4. Isolate intact protoplasts by flotation on sucrose (Section 3.2., step 3).

3.4. Iodoacetate Treatment

Iodoacetate inactivates the cytoplasm of protoplasts. After treatment, they no longer have the ability to grow or divide, and typically die within a week. However, iodoacetate-treated protoplasts fused with untreated protoplasts will continue to survive. Iodoacetate treatment is given just prior to protoplast purification on sucrose (Section 3.2., step 3).

1. Wash protoplasts in CPW salts + 0.48*M* mannitol as in Section 3.2., step 2, and then count live protoplasts with a hemocytometer.

2. Pellet and resuspend protoplasts in 2.5 mM sodium iodoacetate (dissolved in CPW salts + 0.48M mannitol) at a concentration of 1×10^6 protoplasts/mL. Incubate for 30 min at room temperature.
3. Pellet protoplasts, wash once more with 10 mL CPW salts + 0.48M mannitol, and purify by flotation on sucrose as described in Section 3.2., steps 3 and 4.

3.5. Electrofusion of Protoplasts

1. Sterilize the fusion chamber by 15-min immersion in 70% ethanol and air-dry. Just prior to use, rinse the chamber with fusion medium.
2. Individually wash freshly isolated protoplasts (from Section 3.2., step 4) of *N. plumbaginifolia* and *N. tabacum* twice in 10 mL of fusion medium, and resuspend in fusion medium at a concentration of 5×10^5 protoplasts/mL.
3. Mix the two species of protoplasts together at a ratio of 1:1. Load 0.25 mL of this mixture into the fusion chamber and immediately apply an AC-alignment field of 500 kHz, 140 V/cm to the chamber (*see* Notes 9–11).
4. Allow the protoplasts to align for 1.5–3 min, reduce the AC field to 40 V/cm, and apply two DC-square-wave pulses (700 V/cm, 35 ms) 1 s apart. After application of the two DC pulses, gradually reduce the AC-alignment field to 0 V over the next 60 s (*see* Notes 12–14).
5. Gently rock the chamber to dislodge the fused protoplasts, and remove the protoplasts with a Pasteur pipet. Place the protoplasts in a 35 × 15-mm Petri plate. Rinse the chamber with 0.25 mL of fusion medium, and add this rinse to the fused protoplasts. Dilute the protoplasts with 1.0 mL of K$_3$ medium + 0.46M glucose, and culture in the dark at 27°C.

3.6. Induction of Callus Greening and Selection for Kanamycin Resistance in Young Hybrid Calli

This procedure allows screening for hybrid calli based on greening owing to complementation of the Su locus. This screen can also be coupled with selection for kanamycin resistance.

1. Four to 6 d after fusion, dilute the cultures with 0.5 mL of TM-2, and culture in the dark at 27°C. Seven to 10 d after fusion, add 1.0 mL of TM-2, scrape any adhering microcolonies off the bottom with a plastic transfer pipet, and transfer the cultures to a 60 × 15-mm Petri plate.
2. At 11–14 d after fusion, add 3.5 mL of TM-2 and split the cultures.
3. Four days later, dilute the cultures with 3.5 mL of TM-2A, split between two new plates, and transfer cultures to light (20–60 μE/m^2, 27°C). Then, at two successive 4-d intervals, dilute the cultures with 3.5 mL of TM-2A and split. By inclusion of 100 μg/mL kanamycin in the TM-2A medium,

hybrids can be selected for kanamycin resistance simultaneously with the induction of callus greening (*see* Note 15).

4. Three to 4 wk after fusion, centrifuge the microcolonies at 33g for 5 min, resuspend in 5–10 mL of liquid MS medium, and plate on solidified MS medium + 0.2 mg/L BA ± 100 µg/mL kanamycin. Spread microcolonies evenly on the plate and not too densely (<300 colonies/plate). Remove any excess liquid MS medium from the plate. Green colonies should begin to be apparent after 2 wk on this medium (*see* Note 16).

5. Transfer hybrid calli to fresh plates (MS medium + 0.2 mg/L BA ± 100 µg/mL kanamycin) every 4 wk. Shoots will form on this medium.

6. Excise regenerated shoots from hybrid calli and root on half-strength MS with or without 100 µg/mL kanamycin.

3.7. Selection for Kanamycin Resistance by the Agarose Bead Culture Technique

This procedure is based on the agarose bead technique developed by Shillito et al. *(26)*. It provides somewhat quicker selection for kanamycin resistant colonies than that in Section 3.6., and because the hybrid calli are embedded in agarose, it allows accurate counting of the frequency of hybrid colony formation. However, the medium used in this procedure (CM) does not induce greening.

1. Culture the fused protoplasts in the dark at 27°C for 7 d or until they have undergone four to six cell divisions (*see* Note 17).

2. Melt the SeaPlaque agarose medium (CM + 2.4% SeaPlaque agarose + 0.23M mannitol) on a hot plate, and cool until just above its gelling temperature (a 35–40°C water bath is useful here) (*see* Note 18).

3. Scrape the microcolonies off of the bottom of the Petri plates with a plastic transfer pipet, and transfer 2.5-mL aliquots of protoplast suspension to 60 × 15-mm Petri plates. (This may require combining two replicate samples of fused protoplasts.) Add an equal volume of SeaPlaque agarose + CM + mannitol, and mix thoroughly by swirling. Cover the Petri dishes, seal with Parafilm, and let the medium harden for 15 min at 4°C. Culture in the light (20–60 µE/m^2, 27°C) for 7 d.

4. Cut each solidified culture into four wedge-shaped sections, and transfer these agarose wedges into large Petri plates (100 × 15 mm). Add 5 mL of liquid CM media + 0.13M mannitol + 100 µg/mL kanamycin to each culture. Grow these cultures in the light at 27°C for 7 d.

5. Add 5 mL of CM + 100 µg/mL kanamycin to each culture. Culture in the light at 27°C for 7 d.

6. At 28 d postfusion, and weekly thereafter, remove 5 mL from each plate, and replace with 5 mL of fresh CM medium + 100 μg/mL kanamycin.
7. Four to 6 wk after fusion, transfer kanamycin-resistant hybrid calli to solidified CM plates +100 μg/mL kanamycin or to regeneration medium.

3.8. GUS Detection Assay

The following is a quick test for GUS expression that is suitable for assaying large numbers of samples. The samples can be quite small (<100 mg) (*see* Note 19).

1. Excise 2–5 mm² pieces of callus tissue, and place them in individual wells of a microtiter plate. We use Falcon plates, cat. no. 3075 (Becton Dickinson Labware, Lincoln Park, NJ), because these plates are UV translucent.
2. Add 50–100 μL of 100 μ*M* MUG solution into the wells.
3. With wooden applicator sticks (wooden end of a cotton swab), crush the tissue and then incubate the plate at 37°C for 0.5–4 h.
4. View plate under UV light. Positive wells will exhibit blue fluorescence.

4. Notes
4.1. Media

1. Typically, the vitamins and the micronutrients are prepared as 1000X stock solutions and stored as 1-mL aliquots at –20°C. One aliquot of each stock will be used for 1 L of media.
2. The hormones BA and NAA are made up as stock solutions at 0.25 mg/L. The zeatin riboside stock is 1 mg/mL. These hormones need to be dissolved in 1 mL of 1*M* NaOH before the stock solutions are made. Stock solutions of BA and NAA can be stored at 4°C and are stable for 6 wk. Zeatin riboside is less stable, and the stock solution should be stored at –20°C.

4.2. Fusion Chamber

3. Although electrode spacings as narrow as 0.2 mm can be used for fusion of plant protoplasts, the efficiency of fusion is reduced. Electrode spacings <0.2 mm should be avoided for work with plants. Electrode spacings >0.5 mm can also be used. However, as the gap between the electrodes is increased, higher-voltage AC fields have to be used for cell alignment. These higher fields can result in convection within the chamber, which disrupts the necessary cell-to-cell contacts.
4. Ordinary Plexiglas™ cracks after repeated exposure to ethanol; use polycarbonate plastic for construction of fusion chambers.

4.3. Protoplast Isolation

5. It may be necessary to vary both the enzyme concentrations and length of digestion to optimize digestion efficiency in other species. The volume of enzyme solution used depends on the amount of leaf material to be digested. Each Petri plate typically holds 20–30 mL of solution. For a good yield of protoplasts (2–5 × 10^6 intact protoplasts/plate), cut enough leaf strips to cover the surface of the plate.

6. Care should be taken to prevent protoplast lysis during pelleting and resuspension. To reduce the potential for protoplast damage when pipeting, we use 10-mL pipets with their tips cut off to make the bore size larger (2–5 mm). We also pipet the protoplasts slowly (about 1 mL/s) to prevent protoplast lysis. When pelleting, use minimum *g*-forces and times.

7. When recovering protoplasts after flotation on 18% sucrose, it is important to remove as little of the sucrose as possible with the protoplasts. Retention of too much sucrose can make it difficult to pellet the protoplasts at the next step. If this happens, dilute the protoplasts with additional CPW salts + 0.48*M* mannitol, and centrifuge again.

4.4. Irradiation

8. The threshold dose for mitotic inactivation must be checked for each species. Use the lowest dose that completely blocks colony formation in the treated protoplasts. Doses above this threshold increase the frequency of elimination of irradiated chromosomes and result in genetically asymmetric hybrid calli *(18)*.

4.5. Fusion

9. The ratio at which protoplasts are mixed can be varied. An equal ratio between the two species is suggested as a good starting point. If only a few protoplasts of one of the two genotypes is available, we keep the number of protoplasts in the fusion high by adding an excess (2:1 or 3:1) of the more abundant protoplast.

10. Controls should include protoplasts of each type cultured alone (with and without selection), and mixed but unfused protoplasts cultures (with and without selection).

11. AC and DC field parameters are given as field strengths (i.e., V/cm). The actual voltages used will depend on the spacing between electrodes in the chamber. For an electrode gap of 0.5 mm, a 140 V/cm AC field is achieved by applying 7 V to the fusion-chamber electrodes.

12. Cell alignment and fusion can be visually monitored by placing the fusion chamber on an alcohol-sterilized microscope stage. Cover the fusion chamber with a Petri plate top to prevent contamination.

13. If protoplasts do not align in the AC field, first check all electrical contacts. If they still do not align, wash the protoplasts again in fusion medium, and try again. Salt in the medium prevents alignment, and in poor-quality protoplast preparations, cell lysis leads to salt buildup in the medium.

14. The fusion settings we suggest (2 DC pulses of 700 V/cm, 35 μs) are a good starting point. It may be necessary to vary the strength, duration, and number of pulses to maximize protoplast fusions when working with other species or cell types. Increasing the AC alignment voltage also increases fusion efficiency. However increasing the AC voltage damages more protoplasts and results in more multiple fusions, which are counter-productive.

4.6. Selections

15. For the greening protocol, selection for kanamycin resistance is typically begun when the cultures are diluted with TM-2A (14–21 d postfusion). The rationale behind delaying selection so long is that this allows the unfused protoplasts to act as feeder cells or a nurse culture when the cell density of hybrid cells is low.

16. The degree to which greening (complementation of the Su locus) is observed is variable, but improves as the hybrid calli grow older.

17. The timing of the media changes in this protocol depends on the rate of growth of the protoplasts. If the protoplasts are dividing slowly, the cultures should be diluted accordingly. If the culture is diluted too fast, the colonies may begin to die. Excessive browning of the culture starting 24 h after the dilution is a sign of colony death. Selection with kanamycin can be started at the end of the first week, but hybrid calli appear to grow faster if selection is delayed until the end of the second week.

18. For the agarose bead protocol, the use of mannitol as an osmoticum in the culture medium results in more rapid growth of tobacco protoplasts than if sucrose or glucose is used.

19. The GUS assay given here is a rapid screen, but it does result in some false negatives. Positive results (strong blue fluorescence) are reliable. Green calli will give a background of red fluorescence from chlorophyll, but this does not interfere with the identification of calli with moderate to strong GUS activity. A sensitive, quantitative GUS assay was described by Jefferson *(20)* and a detailed protocol is given in Chapter 11.

Acknowledgment

This work was supported by US-Israeli Binational Agricultural Research and Development Fund grant #IS-1631-89.

References

1. Cocking, E. C. (1960) A method for the isolation of plant protoplasts and vacuoles. *Nature* **187**, 962,963.
2. Nagata, T. and Takebe, I. (1971) Plating of isolated tobacco mesophyll protoplasts on agar medium. *Planta* **99**, 12–20.
3. Power, J. B., Cummins, S. E., and Cocking, E. C. (1970) Fusion of isolated plant protoplasts. *Nature* **225**, 1016–1018.
4. Grosser, J. W., Gmitter, F. G., Louzada, E. S., and Chandler, J. L. (1992) Production of somatic hybrid and autotetraploid breeding parents for seedless citrus development. *Hortscience* **27**, 1125–1127.
5. Helgeson, J. P., Haberlach, G. T., Ehlenfeldt, M. K., Hunt, G., Pohlman, J. D., and Austin, S. (1993) Sexual progeny of somatic hybrids between potato and *Solanum brevidens*: potential for use in breeding programs. *Am. Potato J.* **70**, 437–452.
6. Sjödin, C. and Glimelius, K. (1989) Transfer of resistance against *Phoma lingam* to *Brassica napus* by asymmetric somatic hybridization combined with toxin selection. *Theor. Appl. Genet.* **78**, 513–520.
7. Kao, K. N. and Constable, F. (1974) Agglutination and fusion of plant protoplasts by polyethylene glycol. *Can. J. Bot.* **52**, 1603–1606.
8. Scheurich, P. and Zimmermann, U. (1980) Membrane fusion and deformation of red blood cells by electric fields. *Z. Naturforsch.* **35c**, 1081–1085.
9. Zimmermann, U. and Scheurich, P. (1981) High frequency fusion of plant protoplasts by electric fields. *Planta* **151**, 26–32.
10. Bates, G. W., Gaynor, J. J., and Shekhawat, N. S. (1983) The fusion of plant protoplasts by electric fields. *Plant Physiol.* **72**, 1110–1113.
11. Bates, G.W. (1992) Electrofusion of plant protoplasts and the production of somatic hybrids, in *Guide to Electroporation and Electrofusion* (Chang, D. C., Chassy, B. M., Saunders, J. A., and Sowers, A. E., eds.), Academic, San Diego, pp. 249–264.
12. Bates, G. W., Saunders, J. A., and Sowers, A. E. (1987) Electrofusion: principles and applications, in *Cell Fusion* (Sowers, A. E., ed.) Plenum, New York, pp. 367–395.
13. Zimmermann, U., Vienken, J., Pilwat, G., and Arnold, W. M. (1984) Electrofusion of cells: principles and potential for the future, in *Cell Fusion,* CIBA Foundation Symposium 103, Pitman, London, pp. 60–73.
14. Sowers, A. E. (1992) Mechanisms of electroporation and electrofusion, in *Guide to Electroporation and Electrofusion* (Chang, D. C., Chassy, B. M., Saunders, J. A., and Sowers, A. E., eds.) Academic, San Diego, pp. 119–138.
15. Spangenberg, G., Osusky, M., Oilveria, M. M., Freydl, E., Nagel, J., Pais, M. S., and Potrykus, I. (1990) Somatic hybridization by microfusion of defined protoplast pairs in *Nicotiana*: morphological, genetic, and molecular characterization. *Theor. Appl. Genet.* **80**, 577–587.
16. Galun, E. and Aviv, D. (1986) Organelle transfer. *Methods Enzymol.* **118**, 595–611.
17. Bates, G. W. (1992) Molecular analysis of nuclear genes in somatic hybrids. *Physiol. Plantarum* **85**, 308–314.
18. Trick, H., Zelcer, A., and Bates, G. (1994) Chromosome elimination in asymmetric hybrids: effect of gamma dose and time in culture. *Theor. Appl. Genet.* **88**, 965–972.

19. Evans, D. A., Bravo, J. E., Klut, S. A., and Flick, C. E. (1983) Genetic behavior of somatic hybrids in the genus *Nicotiana*: *N. otophora* + *N. tabacum* and *N. sylvestris* + *N. tabacum. Theor. Appl. Genet.* **65,** 93–101.

20. Jefferson, R. A. (1987) Assaying chimeric genes in plants: the GUS gene fusion system. *Plant Mol. Biol. Rep.* **5,** 387–405.

21. Rogers, S. G., Horsch, R. B., and Fraley, R. T. (1986) Gene transfer in plants: production of transformed plants using Ti plasmid vectors. *Methods Enzymol.* **118,** 627–641.

22. Frearson, E. M., Power, J. B., and Cocking, E. C. (1973) The isolation, culture and regeneration of *Petunia* leaf protoplasts. *Dev. Biol.* **33,** 130–137.

23. Nagy, J. I. and Maliga, P. (1976) Callus induction and plant regeneration from mesophyll protoplasts of *Nicotiana sylvestris. Z. Pflanzenphysiol.* **78,** 453–455.

24. Shahin, E. A. (1985) Totipotency of tomato protoplasts. *Theor. Appl. Genet.* **69,** 235–240.

25. Murashige, T. and Skoog, F. (1962) Revised medium for rapid growth and bioassays with tobacco tissue cultures. *Physiol. Plantarum* **15,** 473–497.

26. Shillito, R. D., Paszkowski, J., and Potrykus, I. (1983) Agarose plating and a bead type culture technique enable and stimulate development of protoplasts-derived colonies in a number of plant species. *Plant Cell Reports* **2,** 244–247.

27. Chassy, B. M., Saunders, J. A., and Sowers, A. E. (1992) Pulse generators for electrofusion and electroporation, in *Guide to Electroporation and Electrofusion* (Chang, D. C., Chassy, B. M., Saunders, J. A., and Sowers, A. E., eds.) Academic, San Diego, pp. 555–569.

28. Mischke, J. H., Saunders, J. A., and Owens, L. D. (1986) A versatile low-cost apparatus for cell electrofusion and electro-physiological treatments. *J. Biochem. Biophys. Methods* **13,** 65–75.

29. Zachrisson, A. and Bornman, C. H. (1984) Application of electric field fusion in plant tissue culture. *Physiol. Plantarum* **61,** 314–320.

30. Bates, G. W. (1985) Electrical fusion for optimal formation of protoplast heterokaryons in *Nicotiana. Planta* **165,** 217–224.

CHAPTER 13

Protoplast Electrofusion and Regeneration in Potato

Jianping Cheng and James A. Saunders

1. Introduction

Protoplast fusion and subsequent in vitro plant regeneration, leading to somatic hybridization, offer opportunities for transferring entire genomes from one plant into another, regardless of the interspecific crossing barriers. In contrast to techniques for plant transformation that are aimed at single-gene transfer, protoplast fusion is needed when polygenic traits are concerned, as is frequently encountered in the genetics of higher plants. Several *Solanaceous* species, including potato, have been used with greater success than other higher plant species in somatic hybridization because they are more responsive to the protoplast regeneration process. There are two commonly used procedures to induce cell fusion, namely polyethylene glycol (PEG)-induced protoplast fusion and protoplast electrofusion. These procedures have been the subject of several reviews indicating that electrofusion is generally more efficient *(1–5)*. Electrofusion is superior to PEG-induced protoplast fusion in the following aspects:

1. Simplicity of the fusion process;
2. Less toxicity and less physical damage to the protoplasts;
3. Large fusion volume allowing more protoplasts to be treated; and
4. Fine control of the fusion process with the availability of commercial electrofusion equipment.

From: *Methods in Molecular Biology, Vol. 55: Plant Cell Electroporation and Electrofusion Protocols* Edited by: J. A. Nickoloff Humana Press Inc., Totowa, NJ

The following procedures for protoplast preparation, electrofusion, and regeneration have been used successfully in several potato species, including *Solanum phureja, S. chacoense,* and dihaploid and tetraploid *S. tuberosum,* and they are likely also to be suitable for many other potato species with minor modifications.

2. Materials
2.1. Plants for Protoplast Isolation

Use in vitro grown plants initiated from nodal cuttings of greenhouse- or field-grown plants or from seeds for protoplast preparation. Nodal cuttings with 1 or 2 axillary buds are surface-sterilized by immersion in 20% (v/v) commercial bleach for 10 min (seeds are disinfected in 50% [v/v] commercial bleach for 10 min), then rinsed in sterilized distilled water three times. Bleached ends of the cuttings are excised prior to inserting the lower end of the tissue into the propagation medium (A, Table 1) contained in 25 × 150-mm glass culture tubes or other disposable plastic culture vessels. Place disinfected seeds on the surface of the medium. The cultures are grown at 25°C under cool white fluorescent light (approx 30 μE/m^2/s) with 12 h/d photoperiod. Subculture the plants at monthly intervals.

2.2. Solutions and Media

1. Rinse solution: 0.3M KCl, pH 5.7.
2. Flotation solution: 20% sucrose, pH 5.7.
3. Electrofusion solution: 0.5M manitol. Sterilize all solutions by autoclaving (*see* Note 1).
4. Media: *See* Table 1.

3. Methods
3.1. Protoplast Isolation

1. Isolate protoplasts from young plants (3-wk-old cultures). Excise the upper halves (stems together with leaves) of the plants and immerse in 25 mL of pretreatment solution (G, Table 1) in a sterile 100 × 15-mm Petri dish. Perform this soaking pretreatment at 25°C in the dark for 48 h.
2. Following the preincubation, mince the plant material in a sterile 100 × 15-mm Petri dish containing 1 mL of filter-sterilized cell-wall-digesting solution (B, Table 1) (approx 1 g of plant material/dish) with a pair of surgical scalpels. Add filter-sterilized digesting solution (12 mL) to the minced plant material, seal the dish with parafilm, and wrap in aluminum foil.

Digestion proceeds in the dark at 25°C with gentle shaking (60 rpm) overnight.

3. After completion of digestion, add an equal volume of the rinse solution to the protoplast suspension to lower the density of the digesting solution. Sieve the diluted digested material through a screen (approx 50–80 μm pore diameter) into a 15-mL sterile centrifuge tube and centrifuge at 50g for 5 min.

4. Carefully remove the supernatant, and resuspend the pellet in 10 mL of the flotation solution (*see* Note 2). Carefully layer approx 1 mL of rinse solution on the top of the flotation solution. Centrifuge the tube at 50g for 10 min. Intact protoplasts will float to the interface between the flotation solution and the rinse solution.

5. Collect the protoplasts with a sterilized glass pipet from the sucrose interface, and transfer to a new 15-mL centrifuge tube. Dilute the protoplast suspension with 12 mL of rinse solution, and pellet the protoplasts by centrifugation at 50g for 5 min. It is important not to remove too much (no more than 2 mL) of the suspension liquid while collecting protoplasts from the interface, because too much sucrose solution in the rinse will prevent the protoplasts from pelleting.

6. Remove the supernatant and resuspend the protoplast pellet in 1 mL of 0.5M mannitol for electrofusion. Determine the protoplast density by counting cells under a microscope using a hemocytometer and adjust the protoplast density to 1×10^6 protoplasts/mL for electrofusion (*see* Note 3).

3.2. Electrofusion

1. Perform electrofusion using a square-wave electrofusion apparatus. Cell and protoplast electrofusion can be accomplished using some exponential discharge machines. However, most of these types of machines are designed primarily for electroporation and cannot the function to align cells before the fusion pulse. The use of the square-wave pulse generator also allows successful fusion and electroporation over a much broader range of conditions than the exponential pulse generators (6). A square-wave electroporator that has worked well in our hands is the Electro Cell Manipulator, Model 200 by BTX, Inc. (San Diego, CA). Set alignment voltage meter at 40–80 V/cm field strength and with an alignment duration of 20 s. Set the field strength of the pulse at 1.250 kV/cm with a pulse duration of 60 μs and 1 or 2 consecutive pulses. The second pulse may increase the fusion frequency, but it also decreases the percentage of viable cells. It may be advisable to try one pulse initially to determine if the protoplast population in use can withstand the second pulse.

Table 1
Chemical Composition in 1 L Solution
for Potato Protoplast Preparation, Culture, and Regeneration Media[a]

Component	A	B	C	D	E	F	G
NH_4NO_3	1650	—	—	—	—	1650	—
KNO_3	1900	950	950	950	950	1900	950
KH_2PO_4	170	85	85	85	85	270	85
$MgSO_4 \cdot 7H_2O$	370	185	185	185	185	370	185
$CaCl_2 \cdot 2H_2O$	440	660	660	660	660	440	440
Na_2EDTA	37.3	18.65	18.65	18.65	18.65	37.3	—
$FeSO_4 \cdot 7H_2O$	27.8	13.9	13.9	13.9	13.9	27.8	—
$MnSO_4 \cdot 4H_2O$	22.3	11.15	11.15	11.15	11.15	22.3	—
$ZnSO_4 \cdot 7H_2O$	10.6	5.3	5.3	5.3	5.3	10.6	—
H_3BO_3	6.2	3.1	3.1	3.1	3.1	6.2	—
KI	0.83	0.415	0.415	0.415	0.415	0.83	—
$Na_2MoO_4 \cdot 2H_2O$	0.25	0.125	0.125	0.125	0.125	0.25	—
$CoCl_2 \cdot 6H_2O$	0.025	0.013	0.013	0.013	0.013	0.025	—
$CuSO_4 \cdot 5H_2O$	0.025	0.013	0.013	0.013	0.013	0.025	—
Glycine	2.0	2.0	2.0	2.0	2.0	2.0	2.0
Nicotinic acid	0.5	0.5	0.5	0.5	0.5	0.5	0.5
Pyridoxine	0.5	0.5	0.5	0.5	0.5	0.5	0.5
Thiamine	1.0	10.0	10.0	10.0	10.0	1.0	1.0
Inositol	100.0	100.0	100.0	100.0	100.0	100.0	100.0
Casein hydrolysate	250.0	—	250.0	250.0	250.0	250.0	—
Glutamine	—	—	100.0	100.0	100.0	100.0	—
Adenine sulfate	—	—	—	—	—	100.0	—
Sucrose	20 g	—	10 g	10 g	10 g	10 g	—
Glucose	—	18 g	30 g	30 g	30 g	—	—
Mannitol	—	73 g	—	—	—	—	—
Sorbitol	—	—	50 g	50 g	50 g	30 g	—
PVP-10[b]	—	5 g	—	—	—	—	—
MES	—	1 g	—	—	—	—	—
IAA	—	—	—	—	—	0.01	—
NAA	—	—	1.25	1.25	1.25	—	1.25
2,4-D	—	—	0.25	0.25	0.25	—	—
Zeatin	—	—	1.0	1.0	1.0	3.0	—
BAP	—	—	—	—	—	—	0.5
GA_3	—	—	—	—	—	0.5	—
Agar	7 g	—	—	—	—	—	—
Phytagel	—	—	—	—	2 g	2 g	—
LGT agarose	—	—	—	8 g	—	—	—
pH	5.7	5.7	5.7	5.7	5.7	5.7	5.7

Continued on facing page

2. Gently mix protoplasts in a 1:1 ratio from each fusion partner in a 15-mL centrifuge tube. Transfer 400 µL of protoplast mixture into a 2-mm gap electroporation cuvet (e.g., BTX catalog # 620) and cover with a cuvet lid. Insert the cuvet into the safety chamber at the position for electric discharge, and press the discharge button to start electrofusion (*see* Note 4).

3. After electrofusion, the protoplasts, still contained in the electroporation cuvet, are protected from any physical disturbance in a laminar hood for 30 min to allow protoplast membranes to recover from the electrically induced damage.

4. After the recovery period, gently transfer the protoplasts with a sterile glass pipet into a 15-mL centrifuge tube and centrifuge at 50g for 5 min.

5. Resuspend the protoplast pellet in 0.5 mL of liquid culture medium (C, Table 1).

3.3. Protoplast Culture and Regeneration

1. Culture the electrofusion-treated protoplasts in a semisolid medium by **very gently** mixing 0.5 mL of protoplasts suspended in the liquid culture medium after electrofusion with 0.5 mL of 0.8% (w/v) low-gelling-temperature agarose-embedding medium (D, Table 1) in a 30 × 10-mm Petri dish. The agarose-embedding medium is maintained at 45°C in a water bath before being used to prevent gelling (*see* Notes 5 and 6).

2. Incubate the Petri dishes containing protoplasts in a growth chamber at 25°C in the dark. When protoplast-calli (p-calli) reach approx 1 mm in diameter, they are transferred together with the agarose block onto the top of p-callus growth medium (E, Table 1) in a 100 × 20-mm Petri dish and kept in a growth chamber at 25°C in the dark.

3. P-calli remain on p-callus growth medium without subculture until they are approx 3 mm in diameter. At this size, transfer p-calli onto the shoot-regeneration medium (F, Table 1) in a 100 × 20-mm Petri dish, and grow the callus under an average light intensity of 20 µE/m²/s in a 16 h/d photoperiod. Subculture monthly until shoots regenerate.

---

[a]All media are filter-sterilized, except media A and G are autoclaved. To make a filter-sterilized medium containing a gelling agent (agarose or phytagel), the gelling agent is dissolved separately in distilled water in double concentration, pH adjusted to 5.7, and autoclaved; the other components of the medium are dissolved in another container in distilled water in double concentration, pH adjusted to 5.7, and filter-sterilized; mix the two separately prepared parts 1 to 1 under aseptic conditions when the autoclaved gelling solution cools to 70°C. All units are in milligrams, unless specified.

[b]Abbreviations: PVP-10: Polyvinylpyrrolidone, mol wt 10,000; MES: 2-(*N*-Morpholino) ethanesulfonic acid; NAA: Naphthaleneacetic acid; IAA: Indole-3-acetic acid; 2,4-D: 2,4-Dichlorophenoxyacetic acid; BAP: 6-Benzylaminopurine; GA: Gibberellic acid; LGT agarose: low-gelling-temperature agarose.

4. Cut off regenerated shoots from p-calli when they reach approx 20 mm in length, and plant them into the propagating medium (A, Table 1) in culture tubes. Maintain culture conditions as described earlier for nodal cuttings. These shoots regenerate roots readily in the propagating medium. Supplement the propagating medium with 0.05 mg/L naphthaleneacetic acid (NAA) and 0.05 mg/L gibberellic acid (GA_3) if difficulties in rooting arise.

5. Maintain regenerated plants in vitro under controlled conditions prior to transfer to a greenhouse. To transfer regenerated plants from in vitro culture to a greenhouse, the plants are carefully lifted from agar medium and planted into a garden soil mix in small Jiffy paper garden pots. It is easier to lift the plants from agar medium with their roots intact when the roots are <3 cm in length. Keep the potted plants in a growth chamber for 2 wk under a plastic cover to prevent excessive evaporation. Remove the plastic cover after 2 wk and maintain the plants in the growth chamber without a plastic cover for an additional 2 wk. Transplant the plants and soil from the paper pots into regular garden pots, and transfer to a greenhouse. Shading may be provided during the first 2 wk in a greenhouse.

4. Notes

1. The maintenance of aseptic conditions is vital throughout the entire procedure. To this end, we routinely perform all the procedures in a laminar flow hood, except when the samples are inside sealed vessels. All nonsterile glassware and solutions (unless filter-sterilization is specified) must be autoclaved at 121°C and 20 psi for 20 min. Scalpels must be disinfected by heat with a Bunsen burner or an electric incinerator. Researchers should wear surgical gloves during all operations.

2. Protoplasts are very fragile. Be very gentle when pipeting protoplasts and resuspending a protoplast pellet. Be sure to use only wide-bore pipets and allow the protoplast suspension to drain gently from the pipet with force.

3. Perform electrofusion immediately after protoplast preparation, since protoplasts start regenerating cell walls as soon as they are out of the digesting solution.

4. Use disposable, sterile electroporation cuvets for electrofusion.

5. An efficient system for selecting somatic hybrids is critical for the successful application of protoplast fusion. Early selection at the cellular level or callus level is more efficient than late selection at the plant level. Selectable markers, such as antibiotic or herbicide resistance, can be introduced into fusion partners using gene-transfer techniques. Somatic hybrid selection based on this system has been demonstrated successfully in potato *(7)*. Also, potato mutants resistant to amino acid analogs have been used as the fusion partners to facilitate somatic hybrid selection *(8)*. To avoid the time-

consuming task of producing genetically selectable fusion partners, certain endogenous traits can be used conveniently for somatic hybrid selection or screening. For example, hybrid growth vigor in p-calli may be used in vitro for somatic hybrid selection, and some morphological characteristics can be used for hybrid screening. In our laboratory, hybrid growth vigor in p-calli was used to select somatic hybrids between a Colorado-potato-beetle-resistant clone of *S. chacoense* ($2n = 2x = 24$) and a *S. tuberosum* dihaploid ($2n = 2x = 24$). As a tetraploid hybrid, the p-calli grew faster than the p-calli from either fusion partner. Selection was further defined during shoot regeneration, since p-calli from *S. chacoense* did not regenerate shoots under the defined culture conditions, whereas p-calli from *S. tuberosum* regenerted readily. The production of visible anthocyanin pigments in the shoots and leaves of the *S. chacoense* clone were used as an additional marker for hybrid identification. Those that regenerated shoots with anthocyanins were identified as somatic hybrids, and those that were completely green were regenerates of *S. tuberosum*. This system was very effective, since the putative hybrids were confirmed by peroxidase isozyme electrophoresis, morphological differences, alkaloid profiles, and insect bioassays. Similar characterizations could be used in other potato species and genotypes, and should be considered when choosing fusion partners.

When no selectable or screenable marker is available, a manual system may be considered. In such a system, protoplasts prepared from herbicide-bleached leaves (or suspension cultures and dark-grown calli, as suggested by the present authors) without chlorophyll are stained with fluorescein diacetate (FDA) and then fused with normal chlorophyll-containing protoplasts. After fusion treatments, those protoplasts showing both red and green fluorescence are identified as heterokaryons and hand-picked under a fluorescent microscope equipped with a micromanipulator *(9)*. This system requires low protoplast density culture techniques, which may not be feasible with other potato species.

6. Selected somatic hybrids need to be analyzed to confirm their hybrid nature. Isozyme markers *(10,11)* or DNA markers *(12)* may be used for this purpose. In addition, distinct morphological differences between the two fusion partners are also informative for hybrid identification. Cytological studies may be conducted to investigate the ploidy level and chromosomal numbers of confirmed somatic hybrids.

References

1. Negrutiu, I., de Brouwer, D., Watts, J. W., Sidorov, V. I., Dirks, R., and Jacobs, M. (1986) Fusion of plant protoplast: a study using auxotrophic mutants of *Nicotiana plumbaginifolia,* Viviani. *Theor. Appl. Genet.* **72,** 279–286.

2. Bates, G. W., Saunders, J. A., and Sowers, A. E. (1987) Electrofusion: principles and applications, in *Cell Fusion* (Sowers, A. E., ed.), Plenum, New York, pp. 367–395.

3. Saunders, J. A. and Bates, G. W. (1987) Chemically induced fusion of plant protoplasts, in *Cell Fusion* (Sowers, A. E., ed.), Plenum, New York, pp. 497–520.

4. Fish, N., Karp, A., and Jones, M. G. K. (1988) Production of somatic hybrids by electrofusion in *Solanum. Theor. Appl. Genet.* **76**, 260–266.

5. San, H. L., Vedel, F., Sihachakr, D., and Remy, R. (1990) Morphological and molecular characterization of fertile tetraploid somatic hybrids produced by protoplast, electrofusion and PEG-induced fusion between *Lycopersicon esculentum* Mill. and *Lycopersicon peruvianum* Mill. *Mol. Gen. Genet.* **221**, 17–26.

6. Saunders, J. A., Smith, C. R., and Kaper, J. M. (1989) Effects of electroporation pulse wave on the incorporation of viral RNA into tobacco protoplasts. *Biotechniques* **7(10)**, 1124–1131.

7. Masson, J., Lancelin, D., Bellini, C., Lecerf, M., Guerche, P., and Pelletier, G. (1989) Selection of somatic hybrids between diploid clones of potato (*Solanum tuberosum* L.) transformed by direct gene transfer. *Theor. Appl. Genet.* **78**, 153–159.

8. de Vries, S. E., Jacobsen, E., Jones, M. G. K., Loonen, A. E. H. M., Tempelaar, M. J., and Wijbrandi, J. (1987) Somatic hybridization of amino acid analogue-resistant cell lines of potato (*Solanum tuberosum* L.) by electrofusion. *Theor. Appl. Genet.* **73**, 451–458.

9. Puite, K. J., Roest, S., and Pijnacker, L. P. (1986) Somatic hybrid potato plants after electrofusion of diploid *Solanum tuberosum* and *Solanum phureja. Plant Cell Rep.* **5**, 262–265.

10. Desborough, S. L. (1983) Potato (*Solanum tuberosum* L.), in *Isozymes in Plant Genetics and Breeding,* Part B (Tanksley, S. D. and Orton, T. J., eds.), Elsevier, Amsterdam, pp. 167–188.

11. Shields, C. R., Orton, T. J., and Stuber, C. W. (1983) An outline of general resource needs and procedures for the electrophoretic separation of active enzymes from plant tissue, in *Isozymes in Plant Genetics and Breeding,* Part A (Tanksley, S. D. and Orton, T. J., eds.), Elsevier, Amsterdam, pp. 443–468.

12. Landry, B. S. and Michelmore, R. W. (1987) Methods and applications of restriction fragment length polymorphism analysis to plants, in *Tailoring Genes for Crop Improvement: An Agricultural Perspective.* (Bruening, G., Harada, J., and Hollaender, A., eds.), Plenum, New York, pp. 25–44.

CHAPTER 14

Polymer-Supported Electrofusion of Protoplasts

A Novel Method and a Synergistic Effect

Lei Zhang

1. Introduction
1.1. Fundamental Theory

Effective membrane electroporation is a direct and fast (in the milli-second range) field effect when an external electric field (EF) is applied. The applied field strength E induces the ionic interfacial transmembrane potential difference $\Delta\phi_m$, which represents a contribution E_m $(\Delta\phi_m)$ (transmembrane field strength) to the mean EF force, causing structural rearrangements $\Delta\xi$ in the membrane phase. In brief, the electroporation is a sequence *(1)*:

$$E \rightarrow \Delta\phi_m \rightarrow \Delta\xi \tag{1}$$

Theory guides the relationship:

$$\Delta\phi_m = -1.5\ r\ E\ f(\kappa)\ |\cos\theta| \tag{2}$$

where r is the cell radius and θ is the angle between membrane site and E vector. The conductivity factor $f(\kappa)$ is an explicit function of the geometry of the cell and the conductivity of the solution, the membrane, and the cell interior.

Pore formation is the essential step for fusion of protoplasts. Electroporation is viewed as a transition from a statistical hydrophobic pore to hydrophilic pore, caused by a sudden increase of transmembrane voltage. Generally, an electric field strength $E = 0.2 - 20$ kV/cm and a pulse duration $\Delta t = 0.01 - 100$ ms are used.

From: *Methods in Molecular Biology, Vol. 55: Plant Cell Electroporation and Electrofusion Protocols* Edited by: J. A. Nickoloff Humana Press Inc., Totowa, NJ

Several theories of electroporation and electrofusion have been proposed *(2–16)*. Narrow hydrophobic (HO) pores (P_{HO}) are formed by interfacial ionic polarization and its statistical moment; if the field strength E reaches a critical value E_c at a given pulse duration Δt, broad hydrophilic (HI) pores (P_{HI}) will be formed. The formation of HI pores is slower ($\tau^{-1} = k_{HI} \approx 10^5/s$) compared to HO pores ($\tau^{-1} = k_{HO} \approx 10^6/s$) because the transition $P_{HO} \rightleftharpoons P_{HI}$ involves rotation of lipid molecules in the pore wall *(17)*.

If cells are first brought into contact by mechanical manipulation including sedimentation, chemical treatment, or dielectrophoresis (in which the cells are lined up in chains by applying a low-intensity, high-frequency, oscillating EF), the electroporation can produce a fusion of membranes (electrofusion).

1.2. Chemical Fusion

In contrast to electrofusion, in the presence of neutral polymers, e.g., polyethylene glycol (PEG) and dextran (DX), the processes of chemical fusion take place with monolayer fluctuation after essential changes in the environment of the membrane *(18–22)* with two important steps. First, an adsorption layer is formed on the negatively charged membrane surface, causing dehydration, flattening, or even shrinking. The results are a closer cell contact and hydrophobic interactions. Second, the formation of a depletion layer (free of polymers between the adsorption layer and bulk of solution) yields an osmotic pressure difference between the adsorption layer and the bulk concentration, because of a repulsion between attached and free polymers at a critical concentration.

As a consequence, several properties of a membrane may be changed, including the interfacial tension, the electrophoretic mobility, and hence the ζ potential, the surface dielectric constant, and the surface charge in the presence of adsorbed cations or anions.

In comparison with PEG chemical fusion, the main advantages of electrofusion are high efficiency, speed, nontoxicity, and relative simplicity. Finally, the EF method can be applied to a much wider selection of cell types because of these physical features.

1.3. Polymer-Supported Electrofusion

In the last three years, we have developed a new method, termed polymer-supported electrofusion or chemoelectrofusion, by using barley protoplasts. The main aims were to clarify the behavior of membrane structural and functional changes owing to microenvironment interactions and to determine the optimal conditions for plant protoplast fusion under high pulsation treatment, as well as to build the first stage of a unified model for polymer-supported electrofusion by the combination of an electroporation model with the current hypothesis of polymer fusion (chemical fusion).

These studies demonstrated a synergistic effect of polymer-supported electrofusion. It was hoped that the technique of chemoelectrofusion would allow us to reduce DC-pulse energy and lower chemical concentrations in order to preserve cell viability and to decrease side effects.

From the applied point of view, this method is useful for further exploration of specific substances in medicine and biotechnology. For instance, dextran sulfate (DXS) has a structure very similar to glycosaminoglycans, and has been used as an antiatherosclerotic drug and as a potent agent against HIV infection *(23)*. Serum albumin and DX as solute macromolecules have similar effects on electrofusion of red blood cells to those of hemoglobin *(24)*. PEG is also commonly used in cell fusion experiments, and polyvinylalcohol (PVA) has been used during electrofusion of plant protoplasts to effect plant somatic hybridization *(25)*.

2. Materials

1. Mannitol solution: $0.5M$ mannitol (Merck, Rahway, NJ) in autoclaved distilled $H_2O(dH_2O)$; store at $4°C$.
2. Saccharose solution: $0.6M$ saccharose (Biochemie Kleinmachnow, Berlin, Germany); prepare as for mannitol solution.
3. PEG solution: 2 mM (mol wt 4000, 6000, 40,000, Serva [Heidelberg, Germany] and Ferak [Budapest, Hungary]) in $0.5M$ mannitol solution, pH 6.5, in a small amber glass reagent bottle with a tied stopper.
4. DX solution: DX (mol wt 5000, Kabi [Uppsala, Sweden]; 10,000, Pharmacia [Uppsala, Sweden]; 500,000, Serva); prepare as for PEG solution. Incubate the amber glass briefly at $40°C$ in a water bath for complete dissolution.
5. Serum albumin and blue DX solutions: Serum albumin (mol wt 70,000, Pharma, Dessau) and blue dextran (mol wt 10,900, Pharmacia); prepare as for PEG solution.
6. DXS solution: 0.2 mM DXS (mol wt 500,000, Pharmacia); prepare as for PEG solution.
7. Cellulase solution: 0.5% cellulase (Yakult Biochemical LTD, Osaka, Japan) in $0.6M$ mannitol.

A

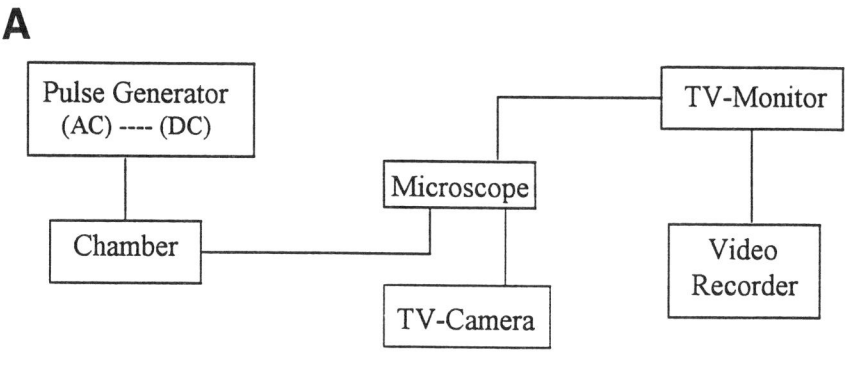

B

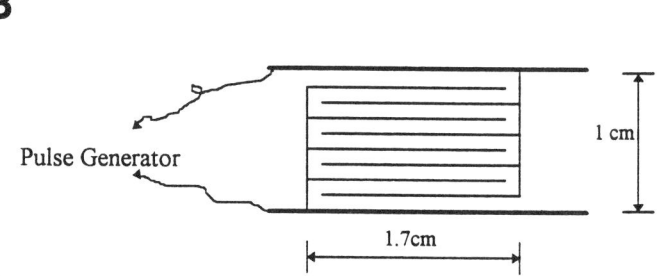

Fig. 1. Schematic of experimental setup. **(A)** The network of equipment for electrofusion. **(B)** A representation of an electrofusion chamber.

8. A pulse generator for producing rectangular pulses (0–100 V pulse height, 0–500 µs duration) in combination with a sine-wave generator (1 MHz) for producing dielectrophoresis.

9. Meander chamber consisting of 60 parallel electrodes separated from each other by distances of 200 µm. The electrode material has a sputtered NiCr/Al of 1.2-µm thickness, which is protected by a 0.035-µm SiO_2 layer. The EF is nonhomogeneous because of the difference between diameter of the protoplast and thickness of the electrode.

10. Microscope (Olympus, Tokyo) linked to a TV monitor and a video recorder (Fig. 1).

3. Methods

3.1. Growth of Barley and Preparation of Protoplasts

1. Plant barley seeds *(Hordeum vulgare)* into nutritious soil. Grow for 9 d under a luminescent UV lamp daily for approx 8 h at a temperature of 25°C.

2. Carefully remove the epidermis of the leaves (approx 20–30) from 9-d-old barley, and cut it into 2-mm pieces. Incubate at 25°C in 5 mL of 5% cellulase solution with gentle shaking for about 1.5–2.0 h. Aspirate the suspension into a centrifuge tube, and centrifuge for 5 min at 850*g*.
3. Aspirate the supernatant, and resuspend protoplasts in 5 mL of 0.6*M* saccharose solution. Above this suspension, carefully pipet 2 mL of 0.6*M* mannitol forming an upper phase. Centrifuge for 5 min at 735*g*. A green ring will appear in the boundary of mannitol and saccharose solutions, which contains protoplasts. Collect protoplasts from the thin ring by very careful aspiration, and split into two or three 0.5-mL centrifuge tubes (*see* Notes 2–5).

3.2. Electrofusion

1. According to Eqs. (3) and (4) to calculate the relative fusion yield F_r.

$$F_r = F_p / F_c \tag{3}$$

$$F = (N_b - N_a)/N_b \tag{4}$$

where F_c is the fusion yield of control measurement (in the absence of polymers), and F_p is the fusion yield of exposure measurement (in the presence of polymers). Calculate F_c and F_p by Eq. (4). N_b is the number of contacts of protoplasts before the pulse, and N_a is the number of nonfused contacts of protoplasts after the pulse.

2. For the control measurement, mix the protoplast suspensions with 0.5*M* mannitol solution at a 1:1 ratio (5–10 µL of each). Then transfer it to the meander chamber, and cover with a small piece of cover glass.
3. Put this chamber under the microscope for observation, and connect it to a pulse generator (AC part). Adjust frequency and voltage to form a "pearl chain" between two electrodes. Count the attachments in various regions of dielectrophoresis alignment.
4. Switch on DC pulse. Apply pulse with suitable strength, duration, and numbers (*see* Note 7), and observe the morphological changes of protoplasts during the fusion process (the contact zone between two membranes increase, the membranes mix, and finally the fused protoplasts become spherical). Count the nonfused attachments in the same regions as before, and calculate the electrofusion yield F_c by Eq. (4).
5. Wash this chamber with distilled H_2O, and dry with a soft clean cloth.
6. For polymer exposure measurement, dilute a polymer stock solution to a certain concentration with 0.5*M* mannitol solution. Mix protoplast suspensions with polymer probes at a ratio of 1:1. Allow a few minutes (ca. 2 min) for interaction between membrane and polymers.

Table 1
Factors Affecting Polymer-Supported Electrofusion

Biological factors	Physical factors	Chemical factors
Plant growth	AC field	Osmosis of medium
Isolation of protoplasts	Frequency and voltage	Molar mass of polymer
Cell type and size	DC pulse	Type of polymer
Physiological state	Strength and duration	Neutral polymer
Viability	Number of pulses	Ampholyte
	Waveform	Polyanion
	Type of fusion chamber	Polycation
	Medium	Concentration
	Ionic strength	Reaction time
	Conductivity	Adsorption
	Permeability	pH
	Electric forces	
	Temperature	

7. Repeat steps 3–5, and calculate F_p by Eq. (4). For polymer-supported electrofusion, use the same electric conditions that yield an F_c value of about 50%.
8. Finally calculate F_r by Eq. (3) (*see* Notes 6–11).

4. Notes
4.1. Biological Properties

1. From the biological, physical, and chemical points of view, there are many variables that can influence the results of polymer-supported electrofusion of protoplasts. To obtain reproducible results, one should be aware of several crucial factors (Table 1). Learning what factors have major or minor effects on the fusion process (as manifested in fusion yield) provides clues to the mechanism of membrane fusion. Our efforts have been directed at measuring the effects of fusion yield as a function of biological and physicochemical parameters. Often the effect of a particular parameter is also dependent on other parameters.
2. Protoplast quality is affected by several factors. First, it is necessary to pay close attention to plant growth. The barley leaves should appear green and healthy; otherwise, the quality of protoplasts will be poor. The viability and permeability of protoplasts may be affected by metabolism. The optimal harvest time is 9–10 d after seeding. Before or after this time, the membranes are not as stable, and hence the fusion yield is not optimal. Second, the processing of protoplasts affects their quality. For instance, if

the incubation time (with cellulase) or the speed of shaking is increased, protoplasts may be damaged. Use caution before and after the second centrifugation (0.6M saccharose and 0.5M mannitol). It is better to lose some protoplasts close to the saccharose phase rather than to aspirate saccharose into the final protoplast suspensions.

3. Protoplast quality is difficult to control. We only use protoplasts of a certain narrow quality range (as judged by microscopic appearance) because these yield reproducible data. For the beginner, it may take practice to become familiar with this technique.

4. Our experimental data have shown that the maximum electrofusion yield is shifted to lower concentrations of polymer if the protoplasts are stored at 4°C for 24 h. Thus, the viability and permeability of protoplasts are optimal for only a few hours after isolation. Also, if protoplast suspensions are held too long at room temperature, then the fusion yield will be reduced.

5. Equation (2) indicates that the effective field strength is related to cell size and cell type. Therefore, the essential step for determining the optimal field parameters is to obtain similar-sized protoplasts by modified centrifugation (e.g., Ficoll gradient). Protoplasts isolated from leaves are generally found to fuse at lower field strengths than protoplasts isolated from roots or suspension cultures, and fusion yield is generally higher for leaf protoplasts under optimal conditions *(26,27)*. It was found that electrofusion yields varied greatly for carrot protoplasts isolated with different commercial enzyme preparations, even though all of the enzymes tested appeared to remove the cell wall completely *(28)*. This indicates that cell-surface preparation can be an important factor in electrofusion.

4.2. Chemical and Physicochemical Properties

6. Osmolarity of fusion medium: Osmotic pressure can be provided by nonionic soluble molecules, which in this procedure is mannitol. Although Ca^{2+} ions (0.5–1.0 mM) have frequently been reported to promote fusion and reduce pulse-induced cell lysis, it is well known that higher ionic strength or electroosmosis can decrease fusion yield and cell viability.

    We found that fusion yield correlated with mannitol concentration. The highest values have been found with 0.5–0.6M mannitol. Below 0.5M, the stability of protoplasts decreases as chloroplasts are lost. Above 0.7M, membrane shrinking and the distribution of chloroplasts are nonhomogeneous. Additionally, the protoplasts become spindle-shaped in the direction of external EF. The optimal condition is 0.55M mannitol, as used for control and polymer-supported experiments. Similar results were reported by Montane and Teissie *(29)*.

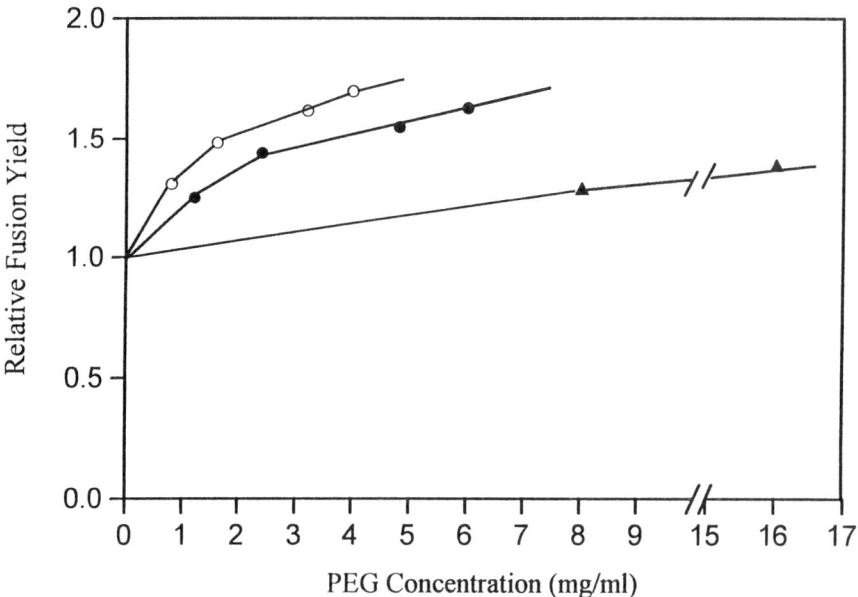

Fig. 2. Influence of molar mass of PEG and its concentration, respectively, on relative fusion yield with PEG 4000 (○), 6000 (●), and 40,000 (▲). Pulse parameters: 1.5 kV/cm, 28 μs, 1 pulse.

7. The molar mass of polymer: Neutral substances, particularly neutral polymers, such as PEG, show an exponential enhancement on fusion yield. Also the lower the molar mass, the higher the fusion yield (Fig. 2); F_r reaches 1.75 with PEG 4000 (4 mg/mL). In comparison with the synergism of PEG, various fractions of DX were tested. For DX 5000 (50 μg/mL) F_r reaches 1.35 as a saturation limit (Fig. 3). DX 5000 results in a higher fusion yield than DX 10,000 and DX 500,000 within the lower range of concentrations.

The neutral polymers accelerate the kinetic process of electrofusion. Under the same EF conditions (AC: 1 MHz, ≤1 kV/cm; DC: 1.5 kV/cm, 28 μs, 1 pulse), protoplasts fuse faster than the control sample by about 1–2 min in the presence of DX 5000 (75 μg/mL) and DXS 500,000 (<250 μg/mL), respectively. Moreover, the fusion yields are enhanced up to 1.35 for DX 5000 and 1.20 for DXS 500,000 (Fig. 3). If the concentration of DX 5000 is higher than 5 mg/mL, the protoplasts are fused during dielectrophoresis without any DC pulse.

This exponential dependence leads to the conclusion that the adsorption of neutral polymers on the membrane plays an important role. For instance,

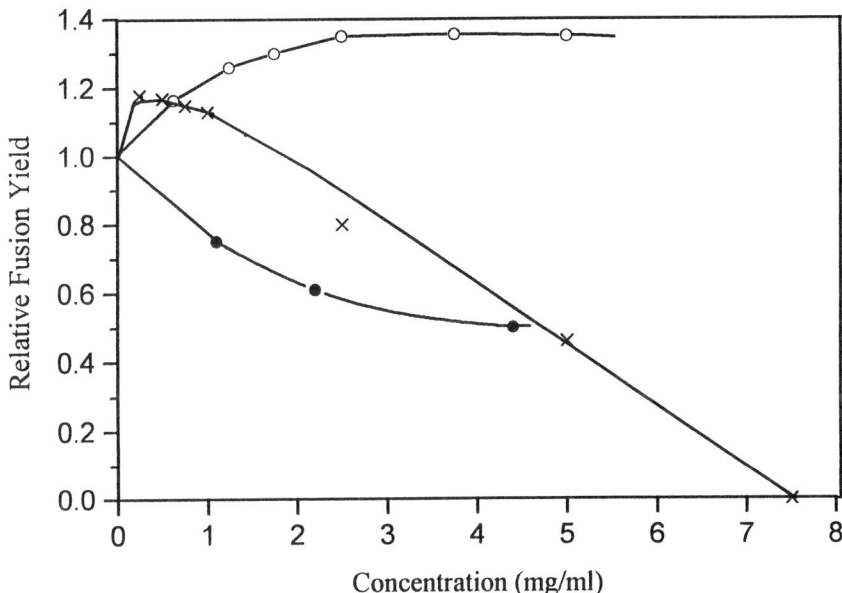

Fig. 3. Influence of concentration of DX 5000 (O), DXS 500,000 (x), and blue DX 10,900 (●), respectively, on relative fusion yield. Pulse parameters: 1.5 kV/cm, 28 μs, 1 pulse (for DX 5000 and DXS 500,000); 1.4 kV/cm, 45 μs, 3 pulses (for blue DX). The DX 5000 concentration was 50 times lower than shown on the axis.

F_r is a function of $[DX]_s$ (the surface concentration of DX), which depends on $[DX]_b$ (the bulk concentration). Consequently, a linearized Freundlich isotherm can be derived by some simplifications ($n < 1$):

$$\log F_r = n \log [DX]_b + \log \beta \tag{4}$$

Equation (4) fits quite well the experimental data for the rising part of the experimental curve, e.g., DX 5000 with $n = 0.54$ and $\log \beta = 0.63$ (Fig. 4).

Neutral polymers at high concentration (>4%) are able to remove the oriented interfacial water dipoles to increase hydrophobicity. Therefore, repulsive forces are reduced. Shrinking and osmotic compression lead to strong contact between membranes, lowering their thermodynamic stability and facilitating fusion *(18,19,30)*. For both PEG and DX, electrofusion yield depends on the molar mass of neutral polymers. With smaller molar mass (PEG 4000, DX 5000), the polymers are more effective than the larger ones (PEG 40,000, DX 500,000) under the EF conditions. Electrofusion yield enhancements in rabbit erythrocyte ghosts by the addition of

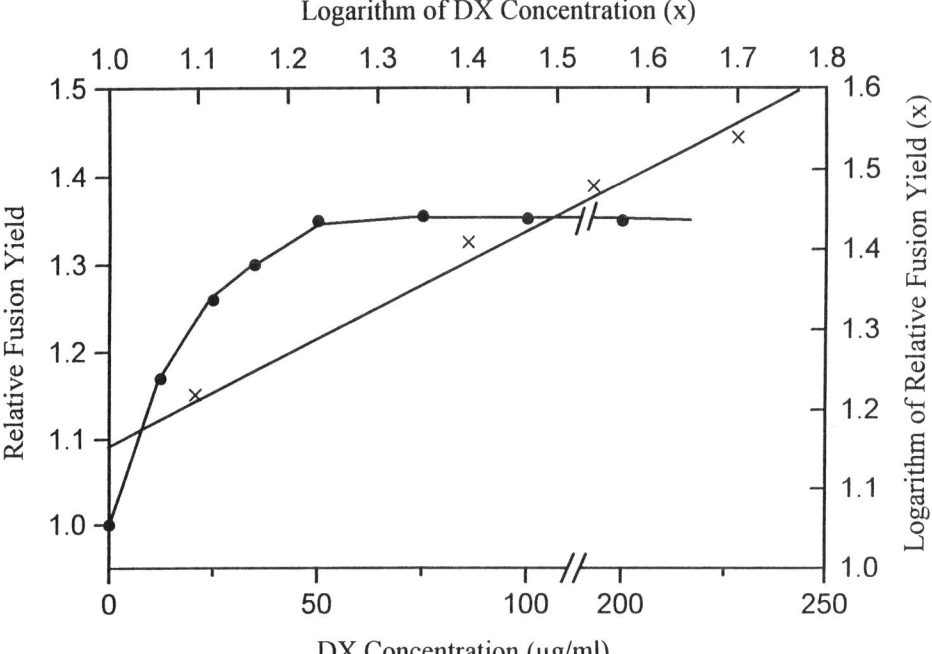

Fig. 4. Influence of the concentration of DX 5000 (●) on the relative fusion yield. The experimental data (x) in the rising part are proportional to the linear curve calculated by Eq. (4). Pulse parameters: 1.5 kV/cm, 28 μs, 1 pulse.

serum albumin or DX 70,000 at low concentrations (0.1–1.0%) were reported to be >2.0. However, this depended on the strength of EF pulse *(24)*. The optimal electroporation conditions have been studied by flow cytometry to determine the amount of uptake of a fluorescent macromolecule (FITC-dextran M_m = 71,000) using intact cells of the yeast *S. cerevisiae (31)*.

8. The charge of polyelectrolytes: The proteinic ampholytes are able to insert their hydrophobic moieties into the lipid layer and shield surface charges of polar head groups. The results obtained from several ampholytes we used have indicated that the electrofusion yield is increased at lower polymer concentrations and is slightly decreased at higher polymer concentrations because of the increasing conductivity (F_r is still >1.0). For instance, F_r = 1.4 with serum albumin at 0.8 mg/mL.

The adsorption behavior of polyelectrolytes can be modified by their positive or negative charge. For polyanions, two types of reactions (Fig. 3) have been detected: First, blue DX decreases electrofusion yield exponentially with increasing concentration owing to the electrostatic repulsion. Second, DXS increases electrofusion yield up to 1.18 only at low concentrations (<1.7 mg/mL). Similar results were seen during chemical fusion of vesicles *(32)*. At low concentrations, the adsorption of the neutral moiety of the sugar chain of DXS compensates for the small conductivity of the solution. Such a positive fusion effect is the result of aggregation induced by DXS adsorption, where the surface of the membrane is covered to a lesser extent. At higher concentrations, strong electrostatic and steric repulsive forces result in a disaggregation on the one hand, and the conductivity of solution becomes higher on the other hand. Thus, electrofusion yield decreases dramatically. This suggests that DXS effects are the result of competition between polymer adsorption and ionic conductivity. The electrofusion phenomenon is very sensitive to the concentrations of this macromolecule.

9. Important factors to note are: (a) the concentrations of polymers and their molar mass, particularly with polyelectrolytes, since at very low concentrations there is a sensitive interaction between cell membranes and macromolecules, and (b) the reaction time (or incubation time) of polymers before applying the pulse. The electrofusion yield of protoplasts depends on the preincubation time with chemicals *(25)*.

10. The ionic strength of the medium is a major factor in its relationship to the conductivity of the cell suspension and in controlling fusion yield. The higher the conductivity of the medium, the lower the fusion yield resulting from dissipation of pulse energy. In addition, this results in a temperature rise of the medium and decreased cell viability.

    Another series of experiments focusing on the modification of electrofusion yield of protoplasts by charged membrane active agents have also been performed in our laboratory *(33)*. One should always take into account the contribution of conductivity in the analysis of data of polyelectrolyte-promoted electrofusion.

11. Polymer-supported electrofusion: A synergism. A combination of electrical and chemical methods provides better conditions than any single method for fast intermingling of adjacent membranes and fusion. This synergistic fusion effect for DX is more effective (by up to 10,000 times) than for PEG concentrations (about 40%) generally used in genetic experiments without pulsation. Why is the relative lower molar mass (M_m) more effective than the higher one? A possible reason is that the surface of shorter polymers comes in contact with the membrane more completely because of a diminished loop formation. The most important force leading to the direct fusion of bilayers is the hydrophobic interaction, which attracts the interiors of membranes *(34)*.

Let us consider a possible mechanism for the detected synergism whereby in the chemical fusion process, adsorption and dehydration increase the hydrophobic interaction energy U_h between two adjacent membranes by decreasing distance from each other *(35)*:

$$U_h = 2\ \sigma \exp\left(-\lambda/\delta\right) \tag{5}$$

where λ is the distance between two membranes, σ is the surface tension of a membrane, and δ is a characteristic decay length of hydrophobic force (1–2 nm). Particularly at very low concentrations of polymers, a bridging mechanism will be favored between two membranes linked by polymer chains.

For electrical fusion, without adsorbed DX, the necessary energy U_p for a large hydrophilic pore (radius $r > 1$ nm) can be calculated with *(11)*:

$$U_p = 2\pi\ (r\ U_e - \sigma_w r^2/2) - 0.5\ \pi\ C\ E^2\ r^2 \tag{6}$$

where U_e is the edge energy of a pore, σ_w is the surface tension of the pore wall ($\sigma_w \approx \sigma$), and C is a capacitance in the pore region of membrane.

In order to describe the results theoretically, it is necessary to combine the derivations of Eqs. (5) and (6) for an unified model. Before application of the pulse, the condition of the cell membrane is modified by adsorption processes of polymers, and this lowers the energy barrier for the formation of hydrophilic pores. It is reasonable to propose that the electrofusion yield generally depends on the extended function:

$$F_r = f(U_h,\ U_p,\ A,\ Q,\ M_m,\ \mathrm{DL}) \tag{7}$$

where A is an adsorption parameter, Q is the surface charge, and DL is the ionic double layer. The value of F_r is significant for synergism; and it is also suitable for the sensitive testing of interactions between substance and membrane (preferably at constant low conductivity of a cell suspension).

The general conclusion regarding polymer-supported electrofusion is that electrical energy can be reduced in the presence of neutral polymers, and the concentration of polymers can be lowered in the presence of EF. This feature provides an advantage for producing high fusion yields of protoplasts and for avoiding unwanted side effects associated with high chemical concentrations or high EF strengths.

References

1. Neumann, E. (1992) Membrane electroporation and direct gene transfer. *Bioelectrochem. Bioenerg.* **28,** 247–268.
2. Kinosita, K. and Tsong, T. Y. (1977) Hemolysis of human erythrocytes by a transient electric field. *Proc. Natl. Acad. Sci. USA* **74,** 1923–1927.

3. Zimmermann, U., Vienken, J., and Pilwat, G. (1980) Development of drug carrier systems: electric field induced effects in cell membranes. *J. Electroanal. Chem.* **116**, 553–574.

4. Neumann, E., Schaefer-Ridder, M., Wang, Y., and Hofschneider, P. H. (1982) Gene transfer into mouse lyoma cells by electroporation in high electric fields. *EMBO J.* **1**, 841–845.

5. Barnett, A. and Weaver, J. C. (1991) Electroporation: a unified, quantitative theory of reversible electrical breakdown and rupture. *Bioelectrochem. Bioenerg.* **25**, 163–182.

6. Sugar, I. P. and Neumann, E. (1984) Stochastic model for electric field-induced membrane pores. *Electroporation Biophys. Chem.* **19**, 211–225.

7. Teissie, J., Knutson, V. P., Tsong, T. Y., and Lane, M. D. (1982) Electric pulse-induced fusion of 3T3 cells in monolayer culture. *Science* **216**, 537,538.

8. Sowers, A. E. (ed.) (1987) *Cell Fusion.* Plenum, New York.

9. Neumann, E., Sowers, A. E., and Jordan, C. A. (eds.) (1989) *Electroporation and Electrofusion Cell Biology.* Plenum, New York.

10. Weaver, J. C. and Barnett, A. (1992) Progress toward a theoretical model for electroporation mechanism: membrane electrical behavior and molecular transport, in *Guide to Electroporation and Electrofusion* (Chang, D. C., Saunders, J. A., Chassy, B. M., and Sowers, A. E., eds.) Academic, San Diego, pp. 91–118.

11. Abidor, I. G., Arakelyan, V. B., Chernomordik, L. V., Chizmadzhev, Y. A., Pastushenko, V. F., and Tarasevich, M. R. (1979) Electric breakdown of bilayer lipid membranes. I. The main experimental facts and their qualitative discussion. *Bioelectrochem. Bioenerg.* **6**, 37–52.

12. Dimitrov, D. S. and Sowers, A. E. (1990) Membrane electroporation—fast molecular exchange by electroosmosis. *Biochem. Biophys. Acta.* **1022**, 381–392.

13. Glaser, R. W., Leikin, S. L., Chernomordik, L. V., Pastushenko, V. F., and Sokirko, A. I. (1988) Reversible electrical breakdown of lipid bilayers: formation and evolution of pores. *Biochem. Biophys. Acta* **940**, 275–287.

14. Neumann, E. (1981) Principles of electric field effects in chemical and biological systems, in *Topics in Bioelectrochemistry and Bioenergetics,* vol. 4. (Milazzo, G., ed.) Wiley, New York, pp. 114–160.

15. Dimitrov, D. S. (1993) Kinetics mechanisms of membrane fusion mediated by electric fields. *Bioelectrochem. Bioenerg.* **32**, 99–124.

16. Sukharev, S. I., Klenchin, V. A., Serov, S. M., Chernomordik, L. V., and Chizmadzhev, Y. A. (1992) Electroporation and electrophoretic DNA transfer into cells. *Biophys. J.* **63**, 1320–1327.

17. Neumann, E., Sprafke, A., Boldt, E., and Wolf, H. (1992) Biophysical considerations of membrane electroporation, in *Guide to Electroporation and Electrofusion* (Chang, D. C., Saunders, J. A., Chassy, B. M., and Sowers, A. E., eds.) Academic, San Diego, pp. 77–90.

18. Van Oss, C. J., Arnold, K., and Coakley, W. T. (1990) Depletion flocculation and depletion stabilization of erythrocytes. *Cell Biophys.* **17**, 1–10.

19. Hui, S. W. and Boni, L. T. (1991) Membrane fusion induced by polyethylene glyco, in *Membrane Fusion* (Wilschut, J. and Hoekstra, D., eds.) Marcel Dekker, New York, pp. 231–253.

20. Arnold, K., Krumbiegel, M., Zschörnig, O., Barthel D., and Ohki, Sh. (1991) Influence of polar polymers on the aggregation and fusion of membranes, in *Cell and Model Membrane Interactions* (Ohki, S., ed.) Plenum, New York, pp. 63–87.

21. Zschörnig, O., Arnold, K., Richter, W., and Ohki, S. (1992) Dextran sulfate-dependent fusion of liposomes containing cationic stearylamine. *Chem. Phys. Lipids* **63**, 15–22.

22. Arnold, K. (1994) Cation-induced vesicle fusion modulated by polymers and proteins, in *Biophysics Handbook on Membranes I: Structure and Conformation* (Sackman, E. and Lipowsky, R., eds.) Elsevier, Amsterdam, pp. 865–916.

23. Mitsuya, H., Looney, D. J., Kuno, S., Ueno, R., Wong-Staal, F., and Broder, S. (1988) Dextran sulfate suppression of viruses in the HIV family: inhibition of virion binding to CD4$^+$ cells. *Science* **240**, 646–649.

24. Sowers, A. E. (1990) Low concentrations of macromolecular solutes significantly affect electrofusion yield in erythrocyte ghosts. *Biochem. Biophys. Acta* **1025**, 247–251.

25. Tsay, Sh., Ernst, R., and Hoffmann, F. (1994) Design, synthesis and application of surface-active chemicals for the promotion of electrofusion of plant protoplasts. Bioelectrochem. *Bioenerg.* **34(2)**, 115–122.

26. Tempelaar, M. J. and Jones, M. G. K. (1985) Fusion characteristics of plant protoplasts in electric fields. *Planta* **165**, 205–216.

27. Tempelaar, M. J., Duyst, A., De Vlas, S. Y., Krol, G., Symmonds, C., and Jones, M. G. K. (1987) Modulation and direction of the electrofusion response in plant protoplasts. *Plant Sci.* **48**, 99–105.

28. Nea, L. J. and Bates, G. W. (1987) Factors affecting protoplasts electrofusion efficiency. *Plant Cell Rep.* **6**, 337–340.

29. Montane, M. H. and Teissie, J. (1992) Electrostimulation of plant protoplast division part I. Experimental results. *Bioelectrochem. Bioenerg.* **29**, 59–70.

30. Rols, M. and Teissie, J. (1990) Implications of membrane interface structural forces in electropermeabilization and fusion. *Bioelectrochem. Bioenerg.* **24**, 101–111.

31. Brown, R. E., Bartoletti, D. C., Harrison, G. I., Gamble, T. R., Bliss, J. G., Powell, K. T., and Weaver, J. C. (1992) Multiple-pulse electroporation: uptake of a macromolecule by individual cells of *Saccharomyces cerevisiae. Bioelectrochem. Bioenerg.* **28**, 235–246.

32. Arnold, K., Ohki, Sh., and Krumbiegel, M. (1990) Interaction of detran sulfate with phospholipid surfaces and liposome aggregation and fusion. *Chem. Phys. Lipids* **55**, 301–307.

33. Zhang, L., Fiedler U., and Berg, H. (1992) Modification of electrofusion of barley protoplasts by membrane-active substances. *Bioelectrochem. Bioenerg.* **27(2)**, 87–96.

34. Helm, C. A. and Israelachvili, J. N. (1993) Forces between phospholipid bilayers and relationship to membrane fusion, in *Methods in Enzymology, vol. 220, Membrane Fusion Techniques* (Düzgünes, N., ed.) Academic, San Diego, pp. 130–143.

35. Israelachvili, J. N. (ed.) (1991) *Intermolecular and Surface Forces.* Academic, San Diego.

Index

Methods in Molecular Biology™ Series

Methods in Molecular Biology™ manuals are available at all medical bookstores. You may also order copies directly from Humana by filling in and mailing or faxing this form to: Humana Press, 999 Riverview Drive, Suite 208, Totowa, NJ 07512 USA, Phone: 201-256-1699/Fax: 201-256-8341.

- [ ] 55. **Plant Cell Electroporation and Electrofusion Protocols**, edited by *Jac A. Nickoloff, 1995* • 0-89603-328-7 • Comb $49.50 (T)
- [ ] 54. **YAC Protocols**, edited by *David Markie, 1995* • 0-89603-313-9 • Comb $69.50 (T)
- [ ] 53. **Yeast Protocols:** *Methods in Cell and Molecular Biology*, edited by *Ivor H. Evans, 1995* • 0-89603-319-8 • Comb $69.50 (T)
- [ ] 52. **Capillary Electrophoresis:** *Principles, Instrumentation, and Applications*, edited by *Kevin D. Altria, 1995* • 0-89603-315-5 • Comb $64.50 (T)
- [ ] 51. **Antibody Engineering Protocols**, edited by *Sudhir Paul, 1995* • 0-89603-275-2 • Comb $69.50
- [ ] 50. **Species Diagnostics Protocols:** *PCR and Other Nucleic Acid Methods*, edited by *Justin P. Clapp, 1995* • 0-89603-323-6 • Comb $69.50 (T)
- [ ] 49. **Plant Gene Transfer and Expression Protocols**, edited by *Heddwyn Jones, 1995* • 0-89603-321-X • Comb $69.50 (T)
- [ ] 48. **Animal Cell Electroporation and Electrofusion Protocols**, edited by *Jac A. Nickoloff, 1995* • 0-89603-304-X • Comb $64.50 (T)
- [ ] 47. **Electroporation Protocols for Microorganisms**, edited by *Jac A. Nickoloff, 1995* • 0-89603-310-4 • Comb $69.50
- [ ] 46. **Diagnostic Bacteriology Protocols**, edited by *Jenny Howard and David M. Whitcombe, 1995* • 0-89603-297-3 • Comb $69.50
- [ ] 45. **Monoclonal Antibody Protocols**, edited by *William C. Davis, 1995* • 0-89603-308-2 • Comb $64.50
- [ ] 44. **Agrobacterium Protocols**, edited by *Kevan M. A. Gartland and Michael R. Davey, 1995* • 0-89603-302-3 • Comb $69.50
- [ ] 43. **In Vitro Toxicity Testing Protocols**, edited by *Sheila O'Hare and Chris K. Atterwill, 1995* • 0-89603-282-5 • Comb $69.50
- [ ] 42. **ELISA:** *Theory and Practice*, by *John R. Crowther, 1995* • 0-89603-279-5 • Comb $59.50
- [ ] 41. **Signal Transduction Protocols**, edited by *David A. Kendall and Stephen J. Hill, 1995* • 0-89603-298-1 • Comb $64.50
- [ ] 40. **Protein Stability and Folding:** *Theory and Practice*, edited by *Bret A. Shirley, 1995* • 0-89603-301-5 • Comb $69.50
- [ ] 39. **Baculovirus Expression Protocols**, edited by *Christopher D. Richardson, 1995* • 0-89603-272-8 • Comb $64.50
- [ ] 38. **Cryopreservation and Freeze-Drying Protocols**, edited by *John G. Day and Mark R. McLellan, 1995* • 0-89603-296-5 • Comb $79.50
- [ ] 37. **In Vitro Transcription and Translation Protocols**, edited by *Martin J. Tymms, 1995* • 0-89603-288-4 • Comb $69.50
- [ ] 36. **Peptide Analysis Protocols**, edited by *Ben M. Dunn and Michael W. Pennington, 1994* • 0-89603-274-4 • Comb $64.50
- [ ] 35. **Peptide Synthesis Protocols**, edited by *Michael W. Pennington and Ben M. Dunn, 1994* • 0-89603-273-6 • Comb $64.50
- [ ] 34. **Immunocytochemical Methods and Protocols**, edited by *Lorette C. Javois, 1994* • 0-89603-285-X • Comb $64.50
- [ ] 33. **In Situ Hybridization Protocols**, edited by *K. H. Andy Choo, 1994* • 0-89603-280-9 • Comb $69.50
- [ ] 32. **Basic Protein and Peptide Protocols**, edited by *John M. Walker, 1994* • 0-89603-269-8 • Comb $59.50 • 0-89603-268-X • Hardcover $89.50
- [ ] 31. **Protocols for Gene Analysis**, edited by *Adrian J. Harwood, 1994* • 0-89603-258-2 • Comb $69.50
- [ ] 30. **DNA–Protein Interactions**, edited by *G. Geoff Kneale, 1994* • 0-89603-256-6 • Paper $64.50
- [ ] 29. **Chromosome Analysis Protocols**, edited by *John R. Gosden, 1994* • 0-89603-243-4 • Comb $69.50 • 0-89603-289-2 • Hardcover $94.50

- [ ] 28. **Protocols for Nucleic Acid Analysis by Nonradioactive Probes**, edited by *Peter G. Isaac, 1994* • 0-89603-254-X • Comb $59.50
- [ ] 27. **Biomembrane Protocols:** *II. Architecture and Function*, edited by *John M. Graham and Joan A. Higgins, 1994* • 0-89603-250-7 • Comb $64.50
- [ ] 26. **Protocols for Oligonucleotide Conjugates:** *Synthesis and Analytical Techniques*, edited by *Sudhir Agrawal, 1994* • 0-89603-252-3 • Comb $64.50
- [ ] 25. **Computer Analysis of Sequence Data:** *Part II*, edited by *Annette M. Griffin and Hugh G. Griffin, 1994* • 0-89603-276-0 • Comb $59.50
- [ ] 24. **Computer Analysis of Sequence Data:** *Part I*, edited by *Annette M. Griffin and Hugh G. Griffin, 1994* • 0-89603-246-9 • Comb $59.50
- [ ] 23. **DNA Sequencing Protocols**, edited by *Hugh G. Griffin and Annette M. Griffin, 1993* • 0-89603-248-5 • Comb $59.50
- [ ] 22. **Microscopy, Optical Spectroscopy, and Macroscopic Techniques**, edited by *Christopher Jones, Barbara Mulloy, and Adrian H. Thomas, 1993* • 0-89603-232-9 • Comb $69.50
- [ ] 21. **Protocols in Molecular Parasitology**, edited by *John E. Hyde, 1993* • 0-89603-239-6 • Comb $69.50
- [ ] 20. **Protocols for Oligonucleotides and Analogs:** *Synthesis and Properties*, edited by *Sudhir Agrawal, 1993* • 0-89603-247-7 • Comb $69.50 • 0-89603-281-7 • Hardcover $89.50
- [ ] 19. **Biomembrane Protocols:** *I. Isolation and Analysis*, edited by *John M. Graham and Joan A. Higgins, 1993* • 0-89603-236-1 • Comb $64.50
- [ ] 18. **Transgenesis Techniques:** *Principles and Protocols*, edited by *David Murphy and David A. Carter, 1993* • 0-89603-245-0 • Comb $69.50
- [ ] 17. **Spectroscopic Methods and Analyses:** *NMR, Mass Spectrometry, and Metalloprotein Techniques*, edited by *Christopher Jones, Barbara Mulloy, and Adrian H. Thomas, 1993* • 0-89603-215-9 • Comb $69.50
- [ ] 16. **Enzymes of Molecular Biology**, edited by *Michael M. Burrell, 1993* • 0-89603-322-8 • Paper $59.50
- [ ] 15. **PCR Protocols:** *Current Methods and Applications*, edited by *Bruce A. White, 1993* • 0-89603-244-2 • Paper $54.50
- [ ] 14. **Glycoprotein Analysis in Biomedicine**, edited by *Elizabeth F. Hounsell, 1993* • 0-89603-226-4 • Comb $64.50
- [ ] 13. **Protocols in Molecular Neurobiology**, edited by *Alan Longstaff and Patricia Revest, 1992* • 0-89603-199-3 • Comb $59.50
- [ ] 12. **Pulsed-Field Gel Electrophoresis:** *Protocols, Methods, and Theories*, edited by *Margit Burmeister and Levy Ulanovsky, 1992* • 0-89603-229-9 • Hardcover $69.50
- [ ] 11. **Practical Protein Chromatography**, edited by *Andrew Kenney and Susan Fowell, 1992* • 0-89603-213-2 • Hardcover $59.50
- [ ] 10. **Immunochemical Protocols**, edited by *Margaret M. Manson, 1992* • 0-89603-270-1 • Comb $69.50
- [ ] 9. **Protocols in Human Molecular Genetics**, edited by *Christopher G. Mathew, 1991* • 0-89603-205-1 • Hardcover $69.50
- [ ] 8. **Practical Molecular Virology:** *Viral Vectors for Gene Expression*, edited by *Mary K. L. Collins, 1991* • 0-89603-191-8 • Paper $54.50
- [ ] 7. **Gene Transfer and Expression Protocols**, edited by *Edward J. Murray, 1991* • 0-89603-178-0 • Hardcover $79.50
- [ ] 6. **Plant Cell and Tissue Culture**, edited by *Jeffrey W. Pollard and John M. Walker, 1990* • 0-89603-161-6 • Comb $69.50
- [ ] 5. **Animal Cell Culture**, edited by *Jeffrey W. Pollard and John M. Walker, 1990* • 0-89603-150-0 • Comb $69.50

Name _____

Department _____

Institution _____

Address _____

City/State/Zip _____

Country _____

Phone #_____ Fax #_____

"T" denotes a tentative price. Prices listed are Humana Press prices, current as of June 1995, and do not reflect the prices at which books will be sold to you by suppliers other than Humana Press. All prices subject to change without notice.
UK, Europe, Middle East, and Africa: Order directly from Chapman & Hall by faxing to: +44-171-522-9623.

Postage & Handling: *USA Prepaid (UPS):* Add $4.00 for the first book and $1.00 for each additional book. *Outside USA* (Surface): Add $5.00 for the first book and $1.50 for each additional book.

- [ ] **My check for $_____ is enclosed** *(Drawn on US funds from a US bank).*
- [ ] Visa   [ ] MasterCard   [ ] American Express

Card # _____

Exp. date _____

Signature _____